Assembling Bus Rapid Transit in the Global South

This book explores the mobile ethnography of Dar es Salaam, where consultants and politicians have planned and implemented a bus rapid transit (BRT) system for two decades. It analyses the dual processes of assembling BRT in the Tanzanian metropolis and establishing BRT as a policy model of and for the Global South.

The book elucidates how policy models are constructed and circulated around the globe and depicts the processes by which they are translated between, and materialise within, specific contexts. It presents the case of BRT to demonstrate how technocrats shape these processes through persuasive work aimed at disseminating and stabilising this transport model, and how local actors influence its adaptation in Dar es Salaam. The book adopts a 'double mobility' approach to show how this ethnography follows travelling consultants, circulating policies and moving buses to explore the fluidity of the BRT model. Linking key debates in policy mobility studies and Science and Technology Studies, enriched with postcolonial perspectives and geographies of transport and infrastructure, it offers new insights into the technopolitics of planning and implementing infrastructure systems.

This book will appeal to academics and students of human geography, transport studies, science and technology studies, and African and development studies interested in the technopolitics of transport planning.

Malve Jacobsen is a geographer specialising in urban and transport studies, Science and Technology Studies, and the relationships between Global South and North. She pursued BA and MA at Humboldt University Berlin and PhD at Goethe University Frankfurt, and currently holds a postdoc position at the University of Bonn.

Transport and Society

Series Editor: John D. Nelson

This series focuses on the impact of transport planning policy and implementation on the wider society and on the participation of the users. It discusses issues such as: gender and public transport, travel for the elderly and disabled, transport boycotts and the civil rights movement etc. Interdisciplinary in scope, linking transport studies with sociology, social welfare, cultural studies and psychology.

Transport Policy
Learning Lessons from History
Edited by Colin Divall, Julian Hine and Colin Pooley

The Mobilities Paradigm
Discourses and Ideologies
Edited by Marcel Endres, Katharina Manderscheid and Christophe Mincke

Ports as Capitalist Spaces
A Critical Analysis of Devolution and Development
Gordon Wilmsmeier and Jason Monios

Non-Motorized Transport Integration into Urban Transport Planning in Africa
Edited by Winnie Mitullah, Marianne Vanderschuren and Meleckidzedeck Khayesi

Assembling Bus Rapid Transit in the Global South
Translating Global Models, Materialising Infrastructure Politics
Malve Jacobsen

For more information about this series, please visit: www.routledge.com/Transport-and-Society/book-series/TSOC

Assembling Bus Rapid Transit in the Global South

Translating Global Models, Materialising Infrastructure Politics

Malve Jacobsen

Routledge
Taylor & Francis Group

LONDON AND NEW YORK

First published 2021
by Routledge
2 Park Square, Milton Park, Abingdon, Oxon OX14 4RN

and by Routledge
52 Vanderbilt Avenue, New York, NY 10017

Routledge is an imprint of the Taylor & Francis Group, an informa business

British Library Cataloguing-in-Publication Data
A catalogue record for this book is available from the British Library

Library of Congress Cataloging-in-Publication Data
A catalog record has been requested for this book

ISBN: 978-0-367-89477-1 (hbk)
ISBN: 978-1-003-01941-1 (ebk)

Typeset in Times New Roman
by codeMantra

Siegelziffer D.30 (Goethe University Frankfurt)
Printed with the kind support of the Hans Böckler Foundation.

Contents

Figures

Acknowledgements

Officially, I am this book's sole author. But inspired by Ian Cook et al.'s notion that we researchers are always collectives, I like to think of this work as by *Jacobsen et al.* Even if the 'et al.' remains an idea, not materialised on the book's cover, the least I can do is acknowledging the co-work, exchange and support of the collective through which this book has emerged.

First, I want to thank all research partners I met in Dar es Salaam, Nairobi and online who were all very open to my research. In Dar es Salaam, people did not get tired of this nosy researcher who asked the same questions again and again. A special thanks goes to the employees of the DART Agency, UDA-RT and VETA. Asanteni wote. I enjoyed spending time with you in cool offices and hot buses. During my time with ITDP in Nairobi, I felt warmly welcomed and accepted in my weird double role. I learned and experienced a lot, not only about global BRT planning, but also about the evil of pedestrian bridges and the good of cycle taxis.

For various kinds of academic and practical support, I thank: Marc Boeckler, Nadine Marquardt and Julia Verne – for their encouragement and inspiration, time and feedback over many years; John Stobart and Jovan Maud – for their sharp eyes on early and late versions of this book; the Hans-Böckler-Stiftung, Gudrun Löhrer and Sven Kesselring – for the scholarship and manifold advice; the Department of Human Geography at Goethe University Frankfurt and the Africa Program at Technical University Darmstadt – for enriching discussions and creative coffee breaks; Alev Coban, Greta Ertelt, Franziska Fay, Lis Hansen, Oliva Kipengele, Frederike Lausch, Lucas Stephan and Christiane Tristl – for sharing their perspectives and reading parts of this book.

Finally, my heartfelt thanks go to my family and friends in and around Flensburg, Berlin, Frankfurt, Bonn and beyond for staying by my side during the long journey on which this book has taken me.

Frankfurt, May 2020

Abbreviations

Tanzanian authorities with both English and Kiswahili names are used in their English version. Organisations whose original names are not in English are translated to English in brackets.

AFCS	automated fare collection system
BRT	bus rapid transit
CEO	chief executive officer
CCM	Chama Cha Mapinduzi (*Revolutionary Party*)
DARCOBOA	Dar es Salaam Commuter Bus Owners Association
DART	Dar es Salaam Rapid Transit
DCC	Dar es Salaam City Council
EU	European Union
GEF	Global Environment Facility
GIZ	Deutsche Gesellschaft für Internationale Zusammenarbeit GmbH (*German Agency for International Cooperation*)
GoT	Government of Tanzania
ISO	International Organization for Standardization
ISP	interim service provider
ITDP	Institute for Transportation and Development Policy
ITS	intelligent transport system
JICA	Japan International Cooperation Agency
km/h	kilometres per hour
LAMATA	Lagos Metropolitan Area Transport Authority
LRT	light rail transit
MRT	mass rapid transit
NBS	National Bureau of Statistics
NGO	non-governmental organisation
NMT	non-motorised transport
pphpd	passengers per hour per direction
PPP	public-private partnership
PRT	private rapid transit
SP	service provider
STA	Sustainable Transport Award

STS	Science and Technology Studies
SUMATRA	Surface and Marine Transport Regulatory Authority
TANROADS	Tanzania National Roads Agency
TOD	transit-oriented development
TRA	Tanzania Revenue Authority
TZS	Tanzanian Shillings
UDA	Shirika la Usafiri Dar es Salaam (*Dar es Salaam Transport Company*)
UDA-RT	Usafiri Dar es Salaam Rapid Transit
UN	Habitat United Nations Human Settlements Programme
UNEP	United Nations Environment Programme
US	United States
USD	US Dollar
VETA	Vocational Education and Training Authority

Practical notes

Anonymisation

I have anonymised all names of individuals in order to protect their personal rights, and in order not to 'latch onto questions of individual responsibility' (Rottenburg 2009). All research partners have been given pseudonyms. References to interviews use the initials of the interlocutors' pseudonyms. However, I have retained the names of public figures, organisations, institutions and places. Full anonymisation would have severely reduced the applicable details and data.

Gender

When making general statements, I use *they/their/them* to include all genders.

I allocated *she/he* to authors from literature and to my research partners according to their name and, if possible, their own description. I decided against using gender-neutral language throughout the work since gender did play a role in certain contexts of my research, even if this role is often hard to specify.

Language

I have translated direct quotes from Kiswahili and German to English to increase readability. The *original* words are provided either directly after the translation or in endnotes. The only Kiswahili expression I have not translated is *daladala*, because doing so would have changed the original meaning and ignored specific context (see Chapter 1).

I have not corrected the language of the interviews because I aim to preserve the authenticity of conversations and situations, and because I do not want to reproduce the idea of a hegemonial 'standard' language (particularly not in postcolonial contexts).

DART timeline

The book focuses on the period from 2014 until 2018.

December 2000	The first line of *Transmilenio* starts operating
January 2003	Enrique Peñalosa visits Dar es Salaam
May 2007	The conceptual design and business plan for DART is completed
	The DART Agency is established
June 2007	Walter Hook and Lloyd Wright publish the first *BRT Planning Guide*
March 2008	BRT Lite opens in Lagos
April 2008	The World Bank approves the loan for DART's Phase 1
August 2009	The first African BRT *Rea Vaya* opens in Johannesburg
September 2010	President Kikwete launches construction works of DART's Phase 1
January 2012	ITDP publishes the first *BRT Standard*
February 2012	Construction works of DART's Phase 1 begin
May 2014	The transaction advisory team publishes the *DART Project Information Memorandum*
June 2014	The transaction advisory team organises the market consultation meeting in Dar es Salaam
	DARCOBOA and UDA merge to UDA-RT
September 2014	The Tanzanian Prime Minister decides for single SP and an ISP of two years
April 2015	UDA, UDA-RT and the DART Agency sign the *ISP Agreement*
August 2015	DART bus drivers training start with two prototype buses
September 2015	ITDP Africa starts officially operating from Nairobi
	UDA-RT's 138 buses and the AFCS arrive in Dar es Salaam

October 2015	Construction works of DART's Phase 1 are fully completed
	Magufuli, the former Minister of Works, becomes Tanzania's new President
January 2016	A new CEO takes over at the DART Agency
	UDA-RT does not get an exemption after a public hearing on the import duties
May 2016	UDA-RT pays import duties for buses and the AFCS to the TRA, and moves to Jangwani depot
	UDA, UDA-RT and the DART Agency sign the *ISP Addendum*
	DART Phase 1 starts operating
June 2016	The Tanzanian government starts first attempt to tender for a second SP
January 2017	President Magufuli officially inaugurates DART's Phase 1
	The World Bank approves the loan for construction works of DART's Phase 3 and 4
May 2017	The Tanzanian government re-advertises the tender for a second SP
June 2017	DART wins the *Sustainable Transport Award*
November 2017	ITDP publishes the new edition of the *BRT Planning Guide*
June 2018	ITDP's webinar on DART goes online
	ITDP's *Mobilize* conference takes place in Dar es Salaam
	The first ITDP-led study tour goes to Dar es Salaam
June 2019	Construction works for Phase 2 start, financed by the African Development Bank
February 2020	The Tanzanian government starts another attempt to tender for a second SP

(To be continued)

1 Introducing bus rapid transit

During the first weeks of the Dar es Salaam Rapid Transit (DART) operations in May 2016, people were shouting *shusha* on their journeys. *Shusha*, meaning 'drop me off' in Kiswahili, is the common term used by passengers to tell the conductor of a Tanzanian minibus that they want to disembark. On some occasions, other passengers laughed at them behaving in a DART bus as if they were in a daladala, the Tanzanian minibus system and prevalent mode of public transport in Dar es Salaam.[1] On other occasions, fellow travellers showed them the stop buttons and explained to them how the new service of bus rapid transit (BRT) differs from daladala. One year later, the service delivery manager of the bus operator posted a photo in a WhatsApp group that serves as a discussion and information platform about operational DART issues. The photo showed a person standing at the roadside and reaching out their arm to flag down a bus, a common gesture to stop a daladala. Another member of the group commented the photo: 'Those villagers' (*Wakijiji* – people who have no idea how urban life works).

In many regards, DART is new – for Tanzania, for Africa and for global networks of transport planning. The new transport system has brought a lot of excitement and change to the city. During the first weeks of service, people marvelled at the buses, and boarded them, wide-eyed, while posting online the pictures they had taken of the buses. They discussed the function of devices and meanings of the signs, the comfort provided by the headroom and seats, and the new view of the city these ninety-centimetre-high buses afford. But why, some asked, do the buses lack air conditioning? And why do drivers stop only at stations, and not along the way when passengers wish to get on or off? For many Dar es Salaam residents, BRT was completely novel. But just a few months after the system's launch, they were using DART coolly, routinely. This new transport system has created new spatialities and temporalities in the urban area. Daladala operators are being rerouted, and residents and businesses need to clear space for the BRT corridor, depots and terminals. Despite transporting just a fraction of the passengers who ride daladala every day, DART has reshaped the city's outward appearance (see Figure 1.1). As the opening example of people attempting to hail a DART bus using daladala techniques illustrates, DART

Figure 1.1 The corridor of DART at Magomeni Junction (Ntevi 09/2016).

has also reshaped the driver-conductor-passenger relationship.[2] Instead of beckoning or shouting 'drop me off' (*shusha*), DART passengers must interact with other 'mediating technologies' (Latour 1994, 1999) to halt a bus. When they need to disembark, they push a red button to open the doors, which is accompanied by a warning signal. At the stations, passengers must pass through turnstiles before they can embark; it is not possible to hail a DART bus from the roadside. Thus, people need to adjust to 'the new art of bus travel' that discloses 'a whole new world', as one Tanzanian journalist described their first rides (Jensen 2016).

Images of DART's physical infrastructure have prevailed over those of daladala, be it in online image searches or on covers of Tanzanian newspapers. The new spatial and temporal effects of DART are repeatedly illustrated in the spectacular shortening of bus journey times: two hours by daladala have become forty minutes by DART. As the journalist Mngodo (2016) describes: 'The blue bus had become a magical capsule that took us through changing worlds within a 40 minutes' drive'. The visually most striking components of DART's physical infrastructure are the sky-blue city buses, the voluminous terminals and stations with cyan roofs, and the long BRT corridor. These material components embodying the promise of high passenger capacity and fast travel are in sharp contrast to daladala buses which are often former Japanese school buses running on unpaved, potholed roads, largely without terminals and stations. Hence, the different appearances of daladala and DART reflect different understandings of urban mobility, whereby people experience DART as a modern, technology-mediated public transport service.

The introduction of this new transport system has led to a far-reaching social, political, economic and ecological transformation. New technologies and people appeared in the city, (re)negotiating practices and forms of governance. For instance, only for the first batch of DART operations, a company with more than eight hundred employees was created and more than twenty kilometres of concrete lanes were constructed. The materiality of the physical infrastructure shapes not only the narratives of BRT. DART's buses are also a symbol of innovation, velocity and comfort since they move a fair bit of people, speed up the city, require and request certain forms of behaviour from their users and drivers. The insertion of a new infrastructure not only creates improvements but also involves disruption, in this case relocations of residents and businesses, including daladala, and obstructions that have continued beyond the construction period. Because a BRT is not a closed system like a metro, constructing a BRT must involve cooperation with and constraints for other modes of transport. Assembling BRT involves the erasure of pre-existing structures, and hence gives rise to tensions that are only partially predictable (Ureta 2015: 13, 93ff.).

This sense of tension between new and old found expression in popular culture. For example, the performer Isack Abeneko uses colourful DART images in his music video about a young villager struggling with the fast pace of urban life (Abeneko 2017). Indeed, newness itself was a challenge, as the CEOs of the two major shareholders expressed to me in interviews. Gabriel Vassanji of the governmental DART Agency said: 'We have one big major challenge: This system is new to us, to all of us – for the government, public and investors' (GV 11/2016). And Heaton Galinoma from the service provider Usafiri Dar es Salaam Rapid Transit (UDA-RT) emphasised: 'The biggest challenge, which we are seeing from today is that we have suddenly entered the world [of] what I have called the "uncharted waters"' (HG 05/2016).

The journey of the BRT model

The story of DART begins in 2002, when the Institute for Transportation and Development Policy (ITDP) and other top global BRT consultants from Latin America and the US approached the Dar es Salaam City Council (DCC). The consultants offered the Tanzanian government the possibility to be part of a pilot project for BRT in African cities, funded by the UN Environment Programme, the Global Environment Facility (GEF) and the World Bank. Subsequently, necessary steps were pursued to realise DART. Funding was allocated; conceptual, technical and operational designs were made; and Tanzanian city officials attended study tours to Bogotá and Miami. In 2007, the Tanzanian government inaugurated a semi-autonomous governmental agency, the DART Agency, which has been the central body coordinating and overseeing the whole process ever since. According to World Bank representatives, DART is planned as a 'top-class BRT' and will be among the 'big league' of global BRT (JK 10/2015). The system is planned in six construction phases and will contain 130 km of BRT corridor along the main road axes. Therewith, DART will be one of the most extensive BRT systems worldwide (DART Agency 2014a: 8). For the Tanzanian branch of the World Bank, DART has national priority because Dar es Salaam is the economic centre of Tanzania: 'If Dar doesn't work, the whole country has a problem' (BN 03/2015).

With the famous *Transmilenio* in Bogotá as its most prominent example, BRT has been circulating as a global policy model. It joins the ranks of other globally circulating best practices, such as the policy model of Business Improvement Districts (McCann and Ward 2010; Ward 2011), urban regeneration projects (González 2011) or harm-reduction drug policies (McCann 2008). Built to gradually replace a minibus system, *Transmilenio* began service in 2000 and has since served as the first comprehensive BRT system (Höhnke 2012: 28). Between 2004 and 2014, the number of cities with BRT grew by almost 400 per cent globally, reaching a total of more than 400 (ITDP 2014; see also Filipe and Maćario 2013: 151). Beginning in the late 1990s, technocrats have been disseminating the BRT model, circulating it particularly in the Global South; from Latin America to Southern and Eastern Asia and to Sub-Saharan Africa (Matsumoto 2006; Mejía-Dugand et al. 2012). Although designs for a bus system with a dedicated median lane and controlled entry points were developed for Chicago as early as 1937, the system *Rede Integrada de Transporte*, launched in 1974 in Curitiba, is commonly considered to be the first BRT (Muñoz and Paget-Seekins 2016). Narratives extolling the benefits of BRT – often created and circulated by BRT proponents and technocrats themselves – emphasise reducing greenhouse gas emissions and car dependency in US cities, or coping with rapid urban growth in Latin American, Asian and African cities. With the label 'high capacity at low cost', BRT is said to work exceedingly well for cities of the Global South (Hensher 2007; Wood 2015a). The global BRT success story continues with the model's arrival on the African continent. In 2008, *BRT Lite* was launched in Lagos; however, most scholars and consultants consider the first line of Johannesburg's *Rea Vaya*, launched in 2009, to be Africa's first BRT. *Rea Vaya* was followed by a series of South African BRT systems that were all modelled on the successful *Transmilenio* (Allen 2013; Behrens et al. 2016a: 11).

Based on an extensive analysis of policy mobilities of BRT between South Africa and South America, Wood (2015c) states that policy circulation is always a political process that is not rational but rather determined by aspirations, ideologies and the positioning of policy makers. Moreover, policy circulation comprises both physical and imagined mobilities to make cities become connected with each other. The perspective of mobile policies draws on the new mobilities paradigm that goes beyond classic transport geography approaches and integrates the mobility of people, things and ideas across spatial and temporal scales so that 'transport is now enmeshed in other forms of circulation and flow' (Schwanen 2016: 132; see also Cresswell 2010, 2012a; Sheller and Urry 2006). Mobilities – understood as more than transport – as well as technologies and infrastructures are politically relevant and integral parts of social and cultural life. However, mobilities research and transport geography are only slowly opening up to postcolonial endeavours and theorising from the Global South. Perspectives and case studies that challenge the 'historical hegemony of predominantly western worldviews, concepts, theories, methods and research practices' (Schwanen 2017: 2) are both necessary and a fruitful enrichment for work on mobilities.

Assembling a travelling model, assembling this book

This book explores what happens when a policy model is implemented in a specific context, shaped, on the one hand, by technocrats and development cooperation agencies that aim to disseminate a transport model globally and, on the other hand, by an internally highly structured network of Tanzanian political-economic elite. It traces the circulation and translation of this travelling model (Behrends et al. 2014; Rottenburg 2009) in order to contribute to discussions on policy mobilities, global planning and infrastructural transition. BRT stands out due to its symbolic language, narrative and storytelling, which sells the transport model as a successful solution for cities of the Global South that not only improves urban transport, but also transforms the whole city, its economy and society. The assembling of BRT in Dar es Salaam demonstrates that BRT is indeed more than transport to the extent that its assembling does not happen without political controversy and deviations from the initial plan. Thereby, this research seeks to contribute to the growing debate on decolonising (geographical) knowledges by introducing BRT and DART – widely distributed transport models in the Global South – to current theorisations of global policies, travelling models and urban transportation.

Despite the lively debates and rapid growth of policy mobilities literature, detailed studies of processes of circulation, translation and mutation remain rare (Healey 2013; Robinson 2011). This research field is still relatively undertheorised, and perspectives drawing on (actor) network approaches or concepts of power configurations are still uncommon. However, such perspectives are necessary to analyse if and how travelling models are embedded into heterogeneous networks and hegemonial structures. Temporalisations, processuality and contingencies (see Li 2007) are particularly important when looking at the translation of BRT because translation implies displacement and mutation. Moreover, up to now mobilising and territorialising policy models have mainly been researched retrospectively; as a consequence, empirical studies have tended to focus only on successfully circulated and implemented policies and disregarded mobilisation attempts that either failed or experienced unexpected transformations.

This book addresses this gap by means of what I call a 'double mobility' approach, for which I have adapted the classic Science and Technology Studies term 'gathering' (Law 2004). Combining methods of global ethnography, the approach allows the researcher to follow an ongoing process in multiple sites while following instances in a specific locale. My method follows mobilities and is itself mobile on two main scales: On a globally relational scale, I followed BRT both to the places where the transport model is created and from whence it is distributed, and to the places where it is materialised and adapted. This also included following BRT narratives virtually in policy papers, newspaper articles, webinars and online chat forums. On a more locally relational scale, I followed the transitional process in Dar es Salaam over space and time. I followed different stages of planning, implementing and operating DART, and I travelled with the buses along the corridor and spent many hours at depots, stations and terminals. I focus not only on broadly discussed global technocrats, international study trips and various policy documents, but also on materialisations and (present) absences of circulated knowledge and narratives, as well as on political controversies of the context in Dar es Salaam. Focusing on the materialisation of global policies, I participated in several DART's central moments of assembling, which were glamorous inauguration events and international visits, as well as moments of controversy and confusion.

This work goes beyond general assumptions of policy mobilities literature and BRT assessments from transport geography by taking more-than-human perspectives on DART and BRT. Following the policies and practices of BRT exposes the mutual shaping of technology and society, as well as the interconnectedness of technology and politics. This means going beyond mere descriptions of heterogeneous networks and technical analyses from a transport planner's point of view. Questions of power and distributed agency are crucial in the analysis. My aim is not to assess whether BRT actually is the only or best option for cities in the Global South with transport challenges and whether DART has been the right choice for Dar es Salaam. Instead, I track globally dominant discourses and their materialisations in Dar es Salaam from different perspectives.

In order to contextualise and conceptualise the processes of infrastructural translation and materialisation in Dar es Salaam, this book has two central objectives. The first is to tell an ethnographic story of transnational transport policies and international BRT planning, thereby tracing the ways in which ideals, experiences and expertise are assembled to an (im)mutable mobile and (presumably) successful BRT model. How do global policies mutate on the move and how fluid is the BRT model in the process of translation, i.e. in the process of transformation, displacement and adaptation? How do global models and their materialisations relate to each other and how are global models assembled in a specific locale? The second objective is to provide a detailed analysis of processes of territorialisation and adaptation of the transport model in Dar es Salaam. Here, I do not analyse in detail the possible long-term effects of DART or socio-spatial and socio-economic changes, such as the structural reconfiguration of neighbourhoods and businesses along the BRT corridor. Rather, my focus is on technopolitical transformations, where I understand technopolitics as 'hybrids of technical systems and political practices that produce(d) new forms of power and agency' (Edwards and Hecht 2010: 619). I ask: How does a globally circulating transport model reshape the transport sector in Dar es Salaam under tensions and controversies? How might the new BRT system, labelled as the 'African BRT', influence other cities in their BRT plans, and does DART impact on the global BRT model? By drawing on vocabulary from Science and Technology Studies, I conceptually contribute to the studies of policy mobilities. I reveal how socio-technical translation – understood as the process of mobilisation and mutation (Callon 1986; Latour 1994) – of models, expertise and materials takes place in contexts of the Global South. I show how adaptation is inherent to translation, and that the subsequent deviation of the model does not risk its existence, since travelling models are fluid and mutable. I realise this framework in my own empirical work of assembling, which is the leading concept of this book.

Many scholars have worked extensively on the circulation of the BRT model. Nonetheless, concerns of translation and context-specific assembling, as well as a focus on power dimensions, remain underrepresented. I consider both cities and policy models as assemblages because they are performative, productive and emergent (Ureta 2015: 11–12). They consist of multiple and relational entities, technologies, politics and actors in diverse configurations, including models, techniques, materials and expertise from elsewhere. Assemblages stand for multiplicity and interdeterminacy, continual transformation and fluidity (McFarlane 2011a: 204, 2011b: 652; Salter 2013: 12). Urban assemblages are not only formed by flows of distributed agency, but they are also shaped by non-coherence, constraints and conditions (Allen 2011; Temenos and McCann 2013: 347). This is because power is not equally distributed (Farías 2011; Ureta 2014). Assemblages are not static; they are in a constant (un)making, (re)arranging, (re)organising, (de)stabilising and fitting together of its heterogeneous elements (see Deleuze and Guattari 2005; Wise 2005). Thinking of an assemblage as both a descriptor and a concept (Anderson and McFarlane 2011; McFarlane and Anderson 2011) is beneficial in gathering circulatory processes of global BRT and DART's assembling in Dar es Salaam.

Global assemblages and BRT

For decades, geographers and anthropologists have argued that the global is constituted locally and vice versa, since nothing purely local or global exists. Thus, neither an absolute territorialisation (understood as a relatively defined and stable state) nor a complete deterritorialisation (understood as a mutable, undefined state) occurs in a globalised, relational world (Brenner 2004: 64; see also DeLanda 2006). Cities and other specific sites are always constituted through their relations with other places and scales, and they are continuously made through various situated practices. Like cities, policies are neither purely global nor purely local. In order not to fall back into the global-local dualism, scholars suggest thinking of mobile policies with concepts such as 'assemblage' and 'topologies'. Policies and policymaking are simultaneously relational and territorial, and mobile policies can be regarded as an outcome of an assembling process through which separate but interconnected sites are territorialised (Prince 2017: 335; Robinson 2013). My overall aim is to think between actors and beyond scales, and thus to think about the local globalness or global localness of circulating policies (Prince 2012; see also Peck and Theodore 2015). Thereby, I follow Roy's (2012) ethnography of circulations (see also Baker and McGuirk 2017). To this end, I chose Dar es Salaam with DART as my starting point and followed its multiple relations to globally acting protagonists and sites.

The conceptualisation of BRT in this book is deeply inspired by Ong and Collier's (2005) work on 'global forms' and 'global assemblages', which offers an approach to conceptualise global phenomena and anthropological problems beyond the local-global dyad (see also Collier 2006; Rabinow 2005). Global forms can be ideas, technologies or policies. They have the capacity to decontextualise and recontextualise as they are contextually unbound:

> Global forms can assimilate themselves to new environments, to code heterogeneous contexts and objects in terms that are amenable to control and valuation. At the same time, the conditions of possibility of this movement are complex. Global forms are limited or delimited by specific technical infrastructures, administrative apparatuses or value regimes, not by the vagaries of a social or cultural field. (Collier and Ong 2005: 11)

When global forms territorialise, global assemblages emerge as actual and specific articulations of these global forms. These articulations occur in specific situations and entail the formation of new relationships that can be material, collective or discursive. The global BRT model can be read as a global form, which territorialises in global assemblages – be it DART in Dar es Salaam, *Transmilenio* in Bogotá or a future BRT system in Nairobi. In order to emphasise the processuality and emergence of DART's assembling, I prefer the verb form 'assembling' over the noun 'assemblage(s)'.[3]

Since BRT has become a global phenomenon, the conceptualisation of global assemblages and global forms provides an appropriate and fruitful perspective for the translation process from global models, policies and ideals to socio-material practices. The global BRT model is significantly shaped by the ITDP, one of this book's protagonists. The NGO has developed the *BRT Standard*, a tool for evaluating and classifying BRT systems, and to make global BRT less mutable within processes of de- and re-contextualisation. Conceptualising the *BRT Standard* as an 'immutable mobile', i.e. objects that hold their shape as they move (Latour 1986, 1987; Law 1986, 2002), raises the question of BRT's (im)mutability: on the one hand, BRT is a heterogeneous assemblage, and on the other hand, the proponents of the model attempt to reduce the model's heterogeneity. Building upon the term 'fluid technology' (De Laet and Mol 2000), I discuss whether technologies need to keep their shape in order to be able to move, or whether they can only be mobilised when they are fluid and flexible, i.e. mutable, translatable and adaptable to different contexts (see also Cook and Ward 2012; McCann and Ward 2013). Hence, to what extent does BRT function as a travelling model (see Behrends et al. 2014; Rottenburg 2009) that is articulated in an institutionalised standard?

As this work shows, DART has undergone various processes of adaptation and resistance. The most striking case of conflict and mutation arose from DART's operational model. Not only is the *BRT Standard* constantly changing through its regular updates and improvements (ITDP 2012, 2016a), but also DART does not appear the way it was planned and deviated from the original model. In interaction with diverse human and nonhuman actors, the BRT system is subjected to ongoing changes; it is continually developing and transforming. By inscribing regulations and experiences into the assemblage, every changing presence and absence of its components has an impact on the DART's shape. Whereas processes of de- and reterritorialisation are mutually constitutive, they are also highly conflictual, since they continually produce, reconfigure and transform the political-economic space (Brenner 2004: 64). Hence, the policies of DART are contingent assemblages full of suspense and tension. Competing interests and forces shape policies, a situation that might lead in some instances to resistance and in other instances to adaptation (Healey 2013: 1510; McCann 2011: 146).

Transport planning is technical and political. It has always been about decreasing travel times, raising the quality of service and reducing fuel consumption. In addition, planning implies persuasion, representing and realising certain ideas and interests more than others. Global consultants might become technocrats that are equipped with best practices and one-sided narratives. BRT planning has become global, and the dissemination of the model has been accelerating. This transport model is an example par excellence of circulating policy models: 'BRT has become the vogue' in urban Africa (Pirie 2014: 136). Global technocrats enable the global form to de- and recontextualise, but so too do various tools of mobilisation such as documents, standards and events. Not only do they provide expertise from the outside, they also become part of political decision making processes by producing 'ostensibly neutral and objective knowledge' (Prince 2017: 338). Building upon Mitchell's (2002) 'modern forms of expertise', Harvey and Knox conceptualise the work of (technical) experts as the 'resolution of

specific problems, which fold the social and the technical together to produce material rearrangements in the name of emancipatory transformation
or "development"' (2015: 8). Expertise is subjective and normative (Collier
2006; Martin and Richards 2001) and thus exercises power. Technology has
multiple meanings in BRT discourses that stand under increasing influence
of technocrats and policy mobilisers. Knowledge and experiences, ideals
and policies are inscribed into technologies (Anderson 2002: 649; Hommels
2005; Martin et al. 2012). The technical features of a BRT system enhance
the alleged stability of the system, separating the new from the existing, re-
defining space and time (Pineda 2010: 137).

ITDP presents BRT at international transport conferences as: 'High quality, high capacity, high speed, customer oriented – not an old bus running in
a bus lane' (ITDP 2018b). Being the most famous BRT proponent globally,
the NGO synthesises BRT as being 'more than bus lanes' (see Figure 1.2).
The organisation describes this transport model as 'the establishment of a
transformed world-class public transport service that is customer oriented
and run on sound economic principles' (ITDP 2017: 90). Along the lines
of 'Think rail, see bus!', BRT systems aim to create a metro-like condition
on the surface by combining the advantages of rail and bus systems (ITDP
2018a). Like rail, they should operate independently of road traffic and congestion through the use of dedicated lanes and off-board fare collection; like

Figure 1.2 DART and BRT described as 'More than bus lanes' (ITDP 2018b).

bus systems, they should be cost-efficient in terms of construction, maintenance and operations. ITDP and its wider network of BRT proponents have successfully leveraged these advantages into a narrative of BRT as 'rapid' in two senses. First, BRT is supposed to be a rapid means of transport, moving people faster than minibuses or individual vehicles. Second, BRTs are claimed to be rapid in their implementation, constructed faster than rail-based systems. Global BRT proponents highlight the system's various advantages, including its high frequency and integrated fare structure:

> BRT is poised to provide significant travel-time savings, which obviously will yield economic benefits to the city, its businesses and its residents. In so doing, it can be expected to shape future growth, attracting new investments and developments along BRT corridors. (TG 12/2016)

It would be hard to find portrayals of BRT without enumerations of auspicious innovative characteristics promising direct improvements – primarily less congestion and more regulation of the public transport sector, as well as grand social, economic and ecological benefits – including in the battle against climate change (ITDP 2018c). Moreover, ITDP shows how a BRT system can even contribute to socio-economic integration and socio-political transformation:

> In Johannesburg, South Africa, the Rea Vaya BRT system is showing the world how high-quality transit can connect poor communities to opportunity and even help heal old wounds of racial segregation.
> (ITDP 2016b)

Thus, BRT must fulfil multiple tasks. The bus system has to offer fast and reliable transport services, reduce greenhouse gas emissions, and lead to social transformation and economic growth.

The African BRT

From the point of view of its proponents, it is vital that BRT succeeds in Dar es Salaam, because the Tanzanian metropolis is supposed to form the basis of an African BRT market for global consultants and international investors. In the early 2000s, ITDP declared that DART would become the first 'full' BRT system in Africa.[4] After South African BRT projects had overtaken DART in the late 2000s, DART operations finally started in 2016, and ITDP continued to use DART as a best practice for promoting BRT across the continent. Keen to spread BRT on the continent, ITDP is consulting an increasing number of African cities. Together with other global BRT actors such as engineering offices and development cooperation agencies, the BRT proponent uses DART to demonstrate that BRT is affordable – and thus possible – in African contexts.[5] Already in 2015, the chief technical advisor of the DART Agency predicted that DART will influence the decision for or against BRT in African cities and beyond:

> Every other city, particularly in Africa, is working somehow – Kampala, Nairobi, Rwanda, Maputo – everybody's working on BRT and everybody's watching what's happening here. So if this fails, it's a disaster. Not just for this country and for the city. [...] We are the guinea pig and everybody will look at Dar. And if it doesn't work here, then we will have big impact on the whole of Africa. And if it works, then everybody will come here.
>
> (HM 03/2015)

The advisor turned out to be right, even though DART continues to have serious operational issues and construction of further corridors is delayed again. DART has created and received attention that goes beyond the city, so that it may well shape the future imaginary of public transport in African metropolises. This attention puts pressure on the global BRT community to promote DART so that it would become a success – or at least so that it would look like a success. As *Transmilenio* has already shown, the narrative of success – constructed by global BRT consultants – is highly performative and becomes globally more present than assessments from the respective city itself about whether the system is presented and perceived as a success or not. This also implies that a BRT system can only reach to the status of best practice once it has global support. The need to succeed had been inscribed into DART from the start.

Point of departure

The DART project is part of a series of strategic programmes to improve urban transport in Dar es Salaam, which are funded by so-called development banks and foreign governmental agencies. The first phase, which is the focal point of this book, was mainly funded by a World Bank loan of 190 million USD (World Bank 2008). The Phase 1 corridor is located on Morogoro Road, the busiest and most congested road in the city and one of the main daladala routes (AM 09/2015; DCC 2007: 4–5). The corridor crosses the city centre and the market area, and connects to the ferry terminals that connect to the Eastern part of the city and to Zanzibar. For Phase 1, the re-settlement of residents and businesses, and the relocation of approximately 1,800 daladala have been completed, and compensations have been paid. Due to complications originating from expropriating house owners, finding suitable construction companies and the need to update the initial conceptual design from 2007 (because of the tremendously increased traffic demand), the project development had been perpetually delayed (JK 10/2015; MN 03/2015). Construction of the physical infrastructure started in 2012. Three years later, buses and equipment for an intelligent transport system (ITS) were ordered, and staff were trained to operate the system. After a delay of several years, bus operations began in May 2016. Because the stake-holders of DART and local operators have not yet come to an agreement for a long-term operational structure, an interim service provider currently operates the system. At the time of writing, an international tendering for a second bus operator is ongoing (Mirondo 2020). Despite its flaws and diffi-culties, DART has received mainly positive feedback from national media and international transport protagonists. Passenger acceptance is reflected in their high numbers, a remarkable and continuously increasing ridership. For the next three construction phases, funding has been secured so that detailed designs and construction works are in the pipeline. Nonetheless, an end of the implementation process is hard to predict.

Dar es Salaam is one of the fastest growing cities worldwide – the per-fect match for BRT. Most BRT consultants use this fact as an opener in talks, papers and interviews. The need for a quick and sustainable public transport solution has been increasing dramatically over the past decade due to the interplay of several developments. Dar es Salaam is undergo-ing an immense process of urbanisation that goes along with urban sprawl, economic growth and an expanding middle class (NBS 2013: 26; Salon and Aligula 2012: 72). Consequently, the need for mobility has been rising, but neither urban roads nor public transport has been sufficiently extended and improved. Private car ownership and the volume of people and goods trav-elling to and through the city have been growing, so that congestion has be-come a serious problem within the urban agglomeration (Melbye et al. 2015; Mkalawa and Haixiao 2014). Accordingly, the Tanzanian country director of the World Bank (2017) sees BRT as vital for Dar es Salaam: 'The BRT is one of the most critical investments that can be made in Dar es Salaam, given the high rate of growth of the city'. From an economic point of view, congestion implies not only wasted time and air pollution but also finan-cial loss. Similar situations and stories about 'the backlog of investment in transport and continued rapid urbanisation' (Pirie 2014: 133) can be found in several cities in the East African region. At the same time, politicians increasingly believe that improved mobility will have a deep impact on the quality of life and will foster sustainable development. This twofold situa-tion has led to an increased engagement of international organisations in the field of urban transportation, and traffic is an increasingly central topic in global discourses. One of the major steps of the past decade was that ur-ban mobility concerns became part of the Sustainable Development Goals (UN-Habitat 2015; see also Jacobsen 2015). In this context, GEF, UN Hab-itat and the German Agency for International Cooperation (GIZ) run the Sustainable Urban Transport Project in various regions worldwide. Like-wise, within the World Bank project Dar es Salaam Metropolitan Develop-ment Project, public transportation and non-motorised transport facilities are central concerns (World Bank 2015a, 2015b).

When the modal share of non-motorised transport (NMT) and public transport is high, urban transport generates lower average carbon emissions than cities in which motorised (private) transport is the primary mode of transport. Hence, cities like Dar es Salaam already have a sustainable transport system that can be expanded: in 2014, only 11 per cent of trips in the city were by private car and 62 per cent by daladala (see Figure 1.3), making daladala the prevalent mode of transport (Nkurunziza et al. 2012: 13; Behrens et al. 2016a). However, politicians see room for improvement, and the DART Agency has declared that 'an efficient, safe, environment-friendly, time-saving, and value-for-money transportation [...] is now a basic need' (DART Agency 2014b). Governments and non-governmental organisations call for mass rapid transit (MRT) solutions like metro, commuter rail, light rail or BRT in African cities. Consultants claim that minibus systems like daladala are unable to transport these growing masses, even though minibus systems are highly efficient: in Dar es Salaam, approximately four million trips are made by daladala every day. By comparison, DART's Phase 1 is expected to provide 300,000 trips daily (DART Agency 2018b).

Many academic talks and policy papers refer to minibuses as 'paratransit' (see Behrens et al. 2016b; Cervero 2013; ITDP 2017). But rather than being 'para' – meaning 'alongside', 'beside' or 'at/to the side of' – minibus systems are the core of public transport in cities like Dar es Salaam. They have fixed routes and networks with transfer points. In 2013, daladala served 98 per cent of road-based public transport (Behrens et al. 2016a: 11; DART Agency 2018b). Daladala carry a negative public image of having unruly drivers

Figure 1.3 Daladala in congestion (author's photo, 03/2015).

who do not care about other road users, unfriendly conductors who try to defraud their passengers of their change, and old buses that provide neither comfort nor safety. International transport researchers and Tanzanian government entities reinforce this negative image of daladala, focusing on the (presumed) lack of professionalism, efficiency and quality (DCC n.d.). In contrast, Rizzo (2011, 2017) depicts the underrepresented perspective of the daladala industry, which must sustain itself without direct public subsidies. He highlights the precarity and harsh working conditions that prevail despite the sector's high efficiency. Daladala drivers and conductors do not have secure work contracts with regulated working hours, and the physical infrastructure, e.g. poorly maintained roads and bus stops without shelters, does not offer daladala the same qualities that DART receives (see also Sohail et al. 2004). However, daladala are not 'outside the system' (Rasmussen 2012); the Tanzanian government has regulated this transport sector since 2004 through issuing route licences, providing coherent route signage on the buses and standardising fares (JR 03/2015; McCormick et al. 2016: 61). Taking these aspects into consideration, daladala do not fit the binary and hierarchical categorisation of (para)transit.

Comparing the images of daladala to the one constructed around DART, these two bus systems appear to be opposing each other. International consultancies, national media and local politicians draw a clear line between minibus and BRT by applying binary word pairs like (un)predictable, (un)comfortable and (un)safe: 'the needed CHANGE shifting from disorganised and unsatisfactory public transport system to the modern organised massive transport system' (DART Agency 2018a). Moreover, daladala is blamed as the main cause of congestion, whereas the rising number of private cars is de facto the bigger concern regarding emissions and road space per capita. This might be because DART was initially planned to primarily replace daladala and attract daladala users, but not to reduce the rising private car use. In an advertising film of DART, the World Bank shows images of a broken daladala being pushed by a conductor while a voice-over comments:

> Congested roadways and heavy traffic make it difficult for businesses to thrive in this port city. So, the government and the World Bank have worked together to create a new Rapid Bus Transit system to speed up rush hour and get these busy Tanzanians where they need to be.
>
> (World Bank 2017)

Also, the Colombian BRT proponent and one of the initiators of *Transmilenio*, Enrique Peñalosa refers to Bogotá's minibus system as the old and to *Transmilenio* as the new system. Setting up BRT as the counterpart to minibus – the new technology against the old – is a common strategy of BRT technocrats to legitimise political decisions (Pineda 2010: 137), whereby the superior new overcomes the deficient old. This tension becomes clear through ITDP's (2016a) systematic distinction between an 'old bus' and a 'high-quality BRT system'. A Tanzanian newspaper reproduces this dichotomy and titles 'Gleaming new buses challenge chaotic old ways in Tanzania', arguing that sweaty daladala passengers enviously look at the clean and shiny new DART buses flying by (The Citizen Tanzania 2016). The lack of appreciating daladala legitimises DART, or in other words: only through emphasising or creating a problem, the need of a solution emerges. DART receives the right to exist through its promise to be the way out of the public transport crisis in Dar es Salaam. However, DART has had troubles to dissociate itself from daladala, i.e. to maintain the image of daladala's counterpart. Apart from the more active role of the state within DART and some (but central) BRT-specific infrastructure components like the dedicated lanes and the control centre, these two bus systems do not differ tremendously.

Many minibus and BRT systems have a lot in common, even if they seem to be quite contrary on the first sight. Also practices of DART and daladala show surprising similarities in various regards, and do not differ as much as technocrats assert. Minibus networks heavily inspire the BRT network regarding routes, as well as the names and locations of stops.[6] Daladala and DART offer high operational frequency with comparable fares and similar operational hours. Nonetheless, there are significant differences between the two bus systems, particularly in respect of operational ideals and practices. First, the conductor's range of duties executed in the daladala system is distributed among several actors of DART including 'mediating technologies' (Latour 1994, 1999) and 'technical devices' (Akrich 1992): DART's buses are equipped with GPS technology and various electronic devices of an ITS; station attendants give information and sell tickets or smart cards, which are scanned by turnstiles that let passengers pass in and out of the station. A second major distinction is the driving style and the enforcement of traffic law. DART drivers broadly apply what they have learned in their training about 'Safety, Comfort and Customer Care' (EM 09/2015). This often-repeated motto finds its expression in a new way of moving through the city. DART drivers generally act according to laws and regulations that have been institutionalised by the Tanzanian state under the guidance of international consultants. Daladala drivers in contrast consider more the actual road conditions than traffic laws – a practice that has turned actual road conditions into unwritten traffic rules. On most roads, the outer lane is primarily used by daladala due to their offensive driving style and frequent stopping, which slows down the average travel speed on that lane and makes it less attractive for other road users. Through these practices, daladala have appropriated a ('dedicated') daladala lane, i.e. a similar kind of traffic priority to DART. Other central distinguishing features considered in the operational models of DART and daladala are ownership, financing and regulation. In contrast to the dispersed private ownership of daladala – whose owners are only partially organised in companies or associations – DART is intended to run under an international public-private partnership (PPP) with two bus operators, a fare collector and fund manager.[7]

A new socio-technical understanding of urban transport

Filled with high promises and ensuing expectations, DART has brought a new understanding of technology and society to the city. The decision to adopt BRT as the solution for Dar es Salaam's traffic problems was never challenged by alternatives. BRT never had to prevail against other MRT options or the idea of fundamentally improving the daladala sector, as the chief technical advisor of the DART Agency elaborates: 'Just about every city in every [African] country has more or less no other option than doing a BRT if they want to remain function. And almost every city is at some stage of thinking about it' (HM 03/2015). The expression of having 'no other option' might be due to the fact that international consultants approached the city officials in Dar es Salaam before the council had actually started thinking about a long-term transition of (public) transport in Dar es Salaam. From then on, transport projects and urban development have been deeply influenced by actors of the development sector who are predominantly BRT proponents. Lloyd Wright and Walter Hook, who promoted BRT to Tanzanian politicians in the early 2000s, describe BRT in the first *BRT Planning Guide* in a typical way that reflects the narratives of global BRT proponents: 'BRT is not just about transporting people. [...] BRT has shown to be an effective catalyst to help transform cities into more liveable and human-friendly environments' (Wright and Hook 2007: ii-iii).

Reasons for choosing BRT and concomitant expectations are similar in various places. Policy actors promise a transport system that is modern and efficient and fulfils the highest standards, be it in Pune or Cairo, Quito or Guangzhou. This strong narrative and the local presence of BRT consultants lead local politicians to the assumption that there is no alternative to BRT. In Dar es Salaam, technocrats agree that public transport is the only way to move in the future – and that BRT is the only affordable public transport solution for Tanzania (BN 03/2015). Also according to the chief technical advisor of the DART Agency, BRT is the only affordable high-capacity system: 'There's no way out' (HM 03/2015). That daladala needs to be replaced by one of the MRT options – metro, commuter rail, light rail or BRT – has also been internalised by the chair of DARCOBOA, the Dar es Salaam Commuter Bus Owners Association (AH 05/2015). Hence, arguments against BRT are rare, almost absent. An often-neglected aspect of BRT construction is resettling residents and relocating businesses due to the massive space necessary for the BRT corridor. The main counterargument is that BRT is limited in its capacity. A BRT system with single lanes can carry approximately 12,000 passengers per hour per direction (pphpd), and a BRT system with express lanes can reach up to 45,000 pphpd. In comparison, metro systems have a capacity of up to 60,000 pphpd (LL 11/2016; see also ITDP 2018a). In the light of ongoing urbanisation, this fact makes BRT only a temporary solution for most places.

Many research partners referred to the city master plan on transport policy and system development for Dar es Salaam as a crucial and binding document for transport planning (JICA 2008). This plan makes DART the principal component in resolving the city's traffic problems. The fundamental transitional process of Dar es Salaam's transport infrastructure goes not only back to the ongoing transition from daladala to DART, but more broadly towards a new image of connectivity and speed. Under John Magufuli, the current President and former Minister of works, the city has received new bridges, flyovers and ring roads. The majority of these large-scale infrastructure projects are funded by the World Bank and designed as PPP (see also Hönke and Cuesta-Fernandez 2017: 2). Understanding (networked) infrastructures as 'the backbone of territorial power' that are at the same time political, social and technical (Boeckler et al. 2018: 13; see also Akrich 1992), DART is an essential part of Magufuli's exercise of technopolitical power. The infrastructural transition changes the face of the city and is dubbed as 'a new principle of order [that] has replaced the principle of chaos' (Mngodo 2016). In consequence, 'Dar is not the city you once knew' (John 2016). Planners, politicians and scholars describe pre-BRT transport in Dar es Salaam as in great need of improvement, like in other target cities of ITDP. In 2006, the leading consultancy working on DART's design presented their ideas to the city and captioned a picture of daladala: 'If nothing is done, there will be chaos!' (Hertel 2008). Policy makers and transport planners legitimise the decision for BRT not only by problematising daladala, but also through glamorous presentations of BRT. For instance, a Word Bank specialist described BRT as a 'no-regret-investment' (BN 03/2015), and an ITDP employee said that BRT was the only solution to meet everyone's needs (ITDP 2018a). Thus, BRT systems are seen as a sustainable, strong, innovative, affordable and pragmatic solution for cities that need to 'ensure that their transit systems keep pace with urban growth' (ITDP 2018c; see also Holzwarth 2012: 32). Particularly cities with a low budget choose BRT because this model promises expediency, low cost and low risk. Another strategic argument against socio-political concerns is that BRT systems can build upon already-existing infrastructures using the road network, as well as the workforce and experience of the minibus sector. Whereby the former was done in Dar es Salaam, the latter did not happen due to political struggles.

Consultants from the World Bank, ITDP and other institutions describe BRT frequently using the verbs 'reducing' (emissions, poverty), 'improving' (life quality, economy, reliability, comfort, safety) and 'saving' (time, money). Narratives differ slightly: while ITDP emphasises BRT's benefits leading to a sustainable city, the World Bank focuses on the economic development. The bank names congestion as a major impediment of Dar es Salaam's economic development since people spend tremendous amounts of time and money on transportation (World Bank 2017). The Tanzanian government aims to guide Dar es Salaam to this sustainable and economically prospering future by formulating a vision and a mission:

> DART Vision: To have a modern public transport system at reasonable cost to the users and yet profitable to operators using high quality capacity buses which meet international standards, environmentally friendly operating on exclusive lanes, at less travelling times. DART Mission: To provide quality, accessible and affordable mass transport system and improve urban mobility for the residents of Dar es Salaam which will subsequently: Enable poverty reduction. Improve living standards. Lead to a sustainable economic growth and act as a pioneer of private and public investment partnership in the sector transport in the city.
>
> (DART Agency 2018a)

Accordingly, DART must meet bigger expectations than improving urban mobility. The system must prove Tanzania's ability to implement economic models according to international ideals. The idea that DART might not only improve transportation but also foster economic growth is based on experiences from elsewhere, like the quote of the World Bank illustrates:

> A significant number of people will have their lives improved. [...] As planned, it's not just the bus shortening and improving a daily commute. It can and will do much more. [...] BRT is excellent for business growth. [...] All these opportunities have been shown to be delivered in other cities as a result of this kind of mass transit system. They can happen as well in Dar es Salaam.
>
> (World Bank 2017)

BRT funders and consultants create this promising image not just for Dar es Salaam, but also for a broader public. BRT proponents pay special attention to other cities that might be interested in orientating towards Dar es Salaam's development, just as the Tanzanian metropolis once was inspired by Bogotá.

Controversial assembling

The global circulation of BRT is a good example of the fact that mobilisation and translation of policies are characterised not only by flows and linearity but also by frictions and multiple temporalities. Several BRT implementations and operations worldwide have faced criticism and deviate from the global story of success. Yet, these dissonant stories are noticeably absent in the BRT narrative. Going into detail and beyond impressive numbers of ridership, even the best practice from Bogotá has been the subject of controversies regarding matters of social exclusion (Casas and Delmelle 2014;

Vecchio 2017). Despite promises made by global BRT proponents, the new transport system is not smoothly replacing the prevalent minibus system. Mobile assemblages are politically contested since they are implicated in the production of power (Cresswell 2012b: 20; Hannam et al. 2006). Following Callon et al. (2009), technoscientific controversies are a response to uncertainties. Thereby, controversies are highly effective rather than destructive because they rediscuss and transform the initial situation. Applying this understanding of technoscientific controversies to the technopolitical dimension of DART, technopolitics can become strategic practices that aim to enforce a certain agenda by embedding policy choices in infrastructure projects.

The Tanzanian state, once largely absent in public transport operations, took on a central role as regulator within this new transport infrastructure. International financiers, consultancies and contractors have also taken part in DART's assembling, establishing ideals and enforcing laws on transportation and international PPPs. These shifting power relations materialise in project delays and questions of participation. Despite the enthusiasm of the early 2000s, the assumption of 'fast policies' (Peck and Theodore 2015) and the promising narrative of BRT's easy adaptability, DART's implementation process has been slow (c.f. ITDP Africa 2020). Indeed, it has mirrored the 'constant, gradual, creeping, at times sluggish and sticky, and at other times loitering instead of prompt and hurried' adoption of BRT in South African cities (Wood 2015b: 2). DART's delays occurred in two phases, each of which reflects conflicts of interest between different stakeholders of the project. Controversies regarding compensation payments and resettlements occurred between the Tanzanian government and international legislation, on the one hand, and between landowners, businesspeople and tenants, on the other (Rizzo 2014, 2017). This book focuses on the second conflict, namely the question of who operates DART, which ultimately links back to concerns of the first struggle between (inter)national governance and locally affected people: who participates in DART, who benefits from the project's PPP arrangement and who is excluded from the new transport system? The controversy has been materialising within the structure of the service provider model, most notably in the allocation of shares and contracts for bus operations, but also for fare collection and fund management.

DART has not met the expectations and requirements of certain political elites, companies and associations (yet). Its conceptual design was controversial, and questions of ownership and power distribution led to creative forms of resistance, foremost from the transport company Shirika la Usafiri Dar es Salaam (UDA). Several of my research partners asserted that UDA was unofficially supported by political elites who enabled UDA to obtain a permanent role in DART operations. A milestone in the process of elites granting favours was the amalgamation of UDA and the bus owners' association DARCOBOA to form a new company called Usafiri Dar es Salaam Rapid Transit (UDA-RT), a company that has since become the uncontested leading player in the city's transport sector. The newly formed company protested the World Bank's condition that it had to collaborate with an international company, adopting an indirect strategy of resistance. It purchased more buses and a more advanced ITS technology at higher capital cost than previously agreed with the DART Agency and the World Bank (c.f. DART Agency 2014a). This massive investment made it impossible for the project leaders to cancel the contract with UDA-RT because they depended on UDA-RT as the biggest transport operator in Dar es Salaam. Moreover, due to ongoing delays and postponements of the system's launch, the project was under serious time pressure to finally start operations.

In sum, DART conveys an ambivalent picture, and the creation of its ostensible success needs to be questioned. On the one hand, politicians and media portray DART as a win-win solution: a driver of economic growth and a promising solution for congestion and air pollution. On the other hand, conflicts and persistent uncertainties are ubiquitous within the process. DART has not only mutated, but it has also been continually downgraded; from the projected first African BRT to first East African BRT, and from a marketed *Gold Standard* to *Silver Standard*. This book reveals that DART is more controversial and messier than its image of success produced and perpetuated by BRT proponents.

Itinerary of this book

This research has taken similar paths and dynamics to the process of BRT circulation and translation itself. It was not only moving straight-ahead, but also back and forth, taking detours, pausing at unexpected points, finding surprising connections and outcomes. The chapters are organised according to processes, theoretical debates and empirical arguments. 'Gathering DART,' Chapter 2, details the ethnographic approach of gathering as an empirical practice to understand ramifying relations (Law 2004). I conceptualise a 'double mobility approach' that is composed of the global circulation of BRT, on the one hand, and of DART operations with buses flowing on the corridor, on the other hand. 'Fluid formations,' Chapter 3, addresses the fluid assembling of the global BRT model by considering the diverse actors involved, as well as the means of mobilisation, translation and standardisation. I argue that the global BRT model is a mutable mobile that nevertheless depends on a certain degree of stabilisation to circulate successfully. In Chapter 4, 'Creating the model of the Global South,' I discuss the technopolitics of global planning and persuasion. Central to the construction of BRT as the model of the Global South are the strong narratives and normativities of BRT's heroic figures and the interdependencies within the global BRT community. Since the politics of persuasion are not free from power dynamics and competing interests, the transport model is also confronted with disagreement and frictions that effect politics in Dar es Salaam and the construction of DART as the African best practice. The follow-ing two chapters explore in detail the controversial and gradual assembling of BRT in Dar es Salaam. Chapter 5, 'Effective operational controversies', deals with the local non-compliance concerning the service provider model of DART, since local operators enforced a reformulation of the operational model and a redistribution of power. I examine how controversial policies might be productive for translating and adopting this travelling model, for both global consultants and local actors. The tensions within the DART project leading to a permanent status of intermittency are discussed in Chapter 6, 'Permanent interim assemblages'. This chapter looks at the grad-ual and multiple temporalities of DART and shows how the controversies have materialised in operational practices. The intermittent nature of the as-sembling of DART manifests in deviations from both the global BRT model and DART-specific plans. Chapter 7, 'To conclude: assembling BRT in the Global South', summarises the main findings of this research by making five main arguments regarding travelling models and their context-specific assemblings. Finally, the chapter provides an outlook on DART's situated-ness between the global narrative of the successful African BRT and local experiences of conflict, deterioration and deviation.

Notes

1 Minibus systems in Tanzania are called 'daladala'. The term can refer to the mode of transport, the service or the buses themselves. In Kiswahili, daladala is both singular and plural.

2 Conductor, or *conda*, is the term used in Tanzania for workers who coordinate daladala journeys. The conductor announces stops and terminus, communi-cates with the driver about where to drop off and pick up passengers, and col-lects fares.

3 For a similar argument, see Law (1993). With reference to Bauman (1992), he claims that 'there is order*ing*, but certainly no ord*er*' since orders are never com-plete (Law 1993: 1; emphasis in original). He emphasises the incompleteness, uncertainty and plurality of any social ordering. Ureta builds up on this and argues that 'all forms of ordering are always emergent and fleeting, existing in one particular performance in a certain temporo-spatiality and disappearing in the next one' (2015: 164).

4 See Chapter 3 for more elaboration on BRT-specific terms like *full BRT* and *BRT Lite*.

5 Research partners often referred to something 'African' of urban transport in Dar es Salaam and Nairobi, and scholars tend to homogenise the continent's diverse urban areas to 'the African city' (see Mbembe and Nuttall 2004, 2008; Parnell and Pieterse 2014, 2015; Robinson 2002; Watson 2009 on 'African Urban-ism'). Myers suggests being careful with this categorisation due to the multiplic-ity within the continent (2011: 104–105). BRT proponents have labelled DART as the 'African BRT model' – a marketing strategy that might convince politi-cians from other African cities to invest in BRT because African cities could (assumingly) learn more from DART than from *Transmilenio* or other BRT systems. However, I did not find anything specific 'African' about BRT, and it turned out that *Transmilenio* could tell me much more about DART than BRT

systems from Johannesburg or Lagos. Being located at the same continent does not necessarily guarantee similarity and comparability.

6 See the matatu network mapped by the project Digital Matatus (C4D LAB 2014), which a consultancy used for its design of future MRT in Nairobi (Gauff Consultants 2014). The consultancy used counts on the traffic demand done by the Digital Matatus team for planning MRT capacity required on Nairobi's main axes (EB 03/2015).

7 For details on DART's operational model, see Chapter 5. So far, a Tanzanian company (UDA-RT) is the single operator of DART.

References

Abeneko, I. 2017. "SALAM (official video)". Accessed at *YouTube* (https://www.youtube.com/watch?v=3lK594i__rg&feature=youtu.be). Published 02.03.2017, retrieved 17.04.2020.

Akrich, M. 1992. "The De-Scription of Technical Objects". In: W. Bijker and J. Law (eds.), *Shaping Technology/Building Society. Studies in Sociotechnical Change.* Cambridge, MA: MIT Press, 205–224.

Allen, H. 2013. "Africa's First Full Rapid Bus System. The Rea Vaya Bus System in Johannesburg, Republic of South Africa". Case Study Prepared for Global Report on Human Settlements 2013. Nairobi: UN-Habitat.

Allen, J. 2011. "Powerful Assemblages?" *Area* 43:2, 154–157.

Anderson, B. and C. McFarlane 2011. "Assemblage and Geography". *Area* 43:2, 124–127.

Anderson, W. 2002. "Postcolonial Technoscience. Introduction". *Social Studies of Science* 32/5–6: 643–658.

Baker, T. and P. McGuirk 2017. "Assemblage Thinking as Methodology. Commitments and Practices for Critical Policy Research". *Territory, Politics, Governance* 5:4, 425–442.

Bauman, Z. 1992. *Intimations of Postmodernity.* London, New York: Routledge.

Behrends, A., S.J. Park and R. Rottenburg 2014. "Travelling Models. Introducing an Analytical Concept to Globalisation Studies". In: ibid. (eds.), *Travelling Models in African Conflict Management. Translating Technologies of Social Ordering.* Leiden: Brill, 1–40.

Behrens, R., D. McCormick and D. Mfinanga 2016a. "An Introduction to Paratransit in Sub-Saharan African Cities". In: ibid. (eds.), *Paratransit in African Cities. Operations, Regulation and Reform.* Oxon, New York: Routledge, 1–25.

Behrens, R., D. McCormick and D. Mfinanga (eds.) 2016b. *Paratransit in Africa Cities. Operations, Regulation and Reform.* Oxon, New York: Routledge.

Boeckler, M., U. Engel and D. Müller-Mahn 2018. "Regimes of Territorialization. Territory, Border and Infrastructure in Africa". In: ibid. (eds.): *Spatial Practices. Territory, Border and Infrastructure in Africa.* Leiden, Boston: Brill, 1–20.

Brenner, N. 2004. *New State Spaces. Urban Governance and the Rescaling of Statehood.* Oxford: Oxford University Press.

C4D LAB, University of Nairobi 2014. "Digital Matatus. The Map. Nairobi Matatu Routes". Accessed at *Digital Matatus* (http://www.digitalmatatus.com/map.html). Retrieved 17.04.2020.

Callon, M., P. Lascoumes and Y. Barthe 2009 (2001). *Acting in an Uncertain World. An Essay on Technical Democracy.* Cambridge, MA: MIT Press.

Callon, M. 1986. "Some Elements of a Sociology of Translation. Domestication of the Scallops and the Fishermen of St Brieuc Bay". In: Law, J. (ed.). *Power, Action and Belief. A New Sociology of Knowledge?* London: Routledge, 196–223.

Casas, I. and E. Delmelle 2014. "Identifying Dimensions of Exclusion from a BRT System in a Developing Country. A Content Analysis Approach". *Journal of Transport Geography* 39, 228–237.

Cervero, R. 2013. "Bus Rapid Transit (BRT). An Efficient and Competitive Mode of Public Transport". Working Paper 2013–01, University of California, Institute of Urban and Regional Development.

Collier, S. and A. Ong 2005. "Global Assemblages, Anthropological Problems". In: A. Ong and S. Collier (eds.), *Global Assemblages. Technology, Politics, and Ethics as Anthropological Problems.* Malden, MA: Blackwell, 3–21.

Collier, S. 2006. "Global Assemblages". *Theory, Culture & Society* 23, 399–401.

Cook, I. and K. Ward 2012. "Conferences, Informational Infrastructures and Mobile Policies. The Process of Getting Sweden 'BID Ready'". *European Urban and Regional Studies* 19:2, 137–152.

Cresswell, T. 2012a. "Mobilities II. Still". *Progress in Human Geography* 36:5, 645–653.

Cresswell, T. 2012b. "Towards a Politics of Mobility". *Environment and Planning D* 28, 17–31.

Cresswell, T. 2010. "Mobilities I. Catching Up". *Progress in Human Geography* 35:4, 550–558.

DART Agency 2018a. "DART. Dar Rapid Transit". Accessed at *DART Agency* (http://www.dart.go.tz/publications/2). Published October 2018, retrieved 11.11.2018.

DART Agency 2018b. "FAQs". Accessed at *DART Agency* (http://dart.go.tz/en/faq/). Published November 2018, retrieved 11.11.2018.

DART Agency 2014a. "Dar Rapid Transit (DART). Project Phase 1. Project Information Memorandum". Final Version, May 2014. Dar es Salaam. (*unpublished*).

DART Agency 2014b. "Welcoming Statement: CEO of DART". Accessed at *DART Agency* (http://dart.go.tz/en/chief-executive-officers-welcoming-statement/). Published June 2014, retrieved 20.07.2016.

DCC 2007. "Consultancy Services for the Conceptual Design of a Long Term Integrated Dar es Salaam BRT System and Detailed Design for Initial Corridor. Annex Volume 6. Physical Design. Chapter 6.2. DART Station Design Report". Final Report, June 2007. (*unpublished*).

DCC no date. "Project. Bus Rapid Transport". Accessed at *DDC* (http://www.dcc.go.tz/projects/bus3rapid3transport.html), retrieved 24.05.2017.

DeLanda, M. 2006. *A New Philosophy of Society. Assemblage Theory and Social Complexity.* London, New York: Continuum.

De Laet, M. and A. Mol 2000. "The Zimbabwe Bush Pump. Mechanics of a Fluid Technology". *Social Studies of Science* 30:2, 225–263.

Deleuze, G. and F. Guattari 2005 (1987). *A Thousand Plateaux. Capitalism and Schizophrenia.* Minneapolis, London: University of Minnesota Press.

Edwards, P. and G. Hecht 2010. "History and the Technopolitics of Identity. The Case of Apartheid South Africa". *Journal of Southern African Studies* 36:3, 619–639.

Farías, I. 2011. "The Politics of Urban Assemblages". *City*, 15:3–3, 365–374.

Filipe, L. and R. Maćario 2013. "A First Glimpse on Policy Packaging for Implementation of BRT Projects". *Research in Transportation Economics* 39, 150–157.

Gauff Consultants 2014. "Nairobi Proposed MRTS/Commuter Rail Network". Prepared for *Republic of Kenya, Ministry of Transport and Infrastructure*. Nairobi. (*unpublished*).

González, S. 2011. "Bilbao and Barcelona 'in Motion'. How Urban Regeneration 'Models' Travel and Mutate in the Global Flows of Policy Tourism". *Urban Studies* 48:7, 1397–1418.

Hannam, K., M. Sheller and J. Urry 2006. "Mobilities, Immobilities and Moorings. Editorial". *Mobilities* 1:1, 1–22.

Harvey, P. and H. Knox 2015. *Roads. An Anthropology of Infrastructure and Expertise*. Ithaca, London: Cornell University Press.

Healey, P. 2013. "Circuits of Knowledge and Techniques. The Transnational Flow of Planning Ideas and Practices". *International Journal of Urban and Regional Research* 37:5, 1510–1526.

Hensher, D. 2007. "Sustainable Transport Systems. Moving Towards a Value for Money and Network-Based Approach and Away from Blind Commitment". *Transport Policy* 14, 98–102.

Hertel, C. 2008. "Development of a BRT system in Dar es Salaam". Cities for Mobility – World Congress, Stuttgart. (*unpublished*).

Höhnke, C. 2012. *Verkehrsgovernance in Megastädten. Die ÖPNV-Reformen in Santiago de Chile und Bogotá*. Stuttgart: Franz Steiner Verlag.

Holzwarth, S. 2012. "Bus Rapid Transit Systems for African Cities". *Trialog* 110, 32–37.

Hommels, A. 2005. "Studying Obduracy in the City. Toward a Productive Fusion between Technology Studies and Urban Studies". *Science, Technology, & Human Values* 30:3, 323–351.

Hönke, J. and I. Cuesta-Fernandez 2017. "A Topolographical Approach to Infrastructure. Political Geography, Topology and the Port of Dar es Salaam". *Environment and Planning D*, 1–20.

ITDP Africa 2020. "Dar es Salaam". Accessed at *ITDP Africa* (https://africa.itdp. org/cities/dar-es-salaam/). Retrieved 17.04.2020.

ITDP 2018a. "BRT Planning 101". Accessed at *ITDP* (https://www.itdp.org/publication/ webinar-brt-planning-101/). Published 02.02.2018, retrieved 17.04.2020.

ITDP 2018b. "BRT Planning 501: Integration". Accessed at *ITDP* https://www. itdp.org/2018/10/09/webinar-brt-501-integration/). Published 09.10.2018, retrieved 17.04.2020.

ITDP 2018c. "What We Do: Public Transport". Accessed at *ITDP* (https://www. itdp.org/what-we-do/public-transport/). Retrieved 11.11.2018.

ITDP 2017. "The Online BRT Planning Guide". 4th Edition. Accessed at *ITDP* (https://brtguide.itdp.org). Published 16.11.2017, retrieved 17.04.2020.

ITDP 2016a. "BRT Standard. 2016 Edition". Accessed at *ITDP* (https://www.itdp. org/2016/06/21/the-brt-standard/). Published 21.06.2016, retrieved 17.04.2020.

ITDP 2016b. "How Did Johannesburg, South Africa Create a More Inclusive City?" Accessed at *Campaign Archive* (https://us7.campaign-archive.com/?u=0b5e10eda9 e3afdb7eceb76f6&id=1883bf6aa6&e=2864d0af9a). Published 07.12.2016, retrieved 17.04.2020.

ITDP 2014. "A Global Rise in Bus Rapid Transit: 2004–2014". Accessed at *ITDP* (https://www.itdp.org/global-rise-bus-rapid-transit-2004-2014/). Published 17.11.2014, retrieved 17.04.2020.

ITDP 2012. "BRT Standard. Version 1.0". Accessed at *ITDP* (https://www.itdp.org/ brt-standard-scores/), retrieved 17.04.2020.

Jacobsen, M. 2015. "Schnellbusse auf der Überholspur. Dar es Salaam Rapid Transit als Wegbereiter nachhaltiger urbaner Mobilität?" *RaumPlanung* 182:6, 66–73.

Jensen, A. 2016. "Dar Rapid Transit and the New Art of Bus Travel". Accessed at *The Citizen Tanzania* (http://www.thecitizen.co.tz/magazine/soundliving/1843780-3352662-22de9h/index.html). Published 21.08.2016, retrieved 17.04.2020.

JICA 2008. "Dar es Salaam Transport Policy and System Development Master Plan. Final Report". Tokyo. (*unpublished*).

John, M. 2016. "No, This Is Not the Dar City You Once Knew". Accessed at *The Citizen Tanzania* (http://www.thecitizen.co.tz/News/No--this-is-not-the-Dar-City-you-once-knew/-/1840340/3162872/-/7ejg7a/-/index.html). Published 17.04.2016, retrieved 17.04.2020.

Latour, B. 1999. *Pandora's Hope. Essays on the Reality of Science Studies*. Cambridge, MA: Harvard University Press.

Latour, B. 1994: "On Technical Mediation". *Common Knowledge* 3:2, 29–64.

Latour, B. 1987. *Science in Action. How to Follow Scientists and Engineers through Society*. Cambridge: Harvard University Press.

Latour, B. 1986: "Visualisation and Cognition. Drawing Things Together". *Knowledge and Society – Studies in the Sociology of Culture Past and Present* 6, 1–40.

Law, J. 2004. *After Method. Mess in Social Science Research*. London, New York: Routledge.

Law, J. 2002: "Objects and Spaces". *Theory Culture & Society* 19:5/6, 91–105.

Law, J. 1993. *Organizing Modernity*. Oxford, Cambridge: Blackwell.

Law, J. 1986. "On the Methods of Long Distance Control. Vessels, Navigation and the Portuguese Route to India". In: ibid. (ed.), *Power, Action and Belief: A New Sociology of Knowledge?* London: Routledge, 234–263.

Li, T. 2007. *The Will to Improve. Governmentality, Development, and the Practice of Politics*. Durham, London: Duke University Press.

Martin, B., P. Nightingale and A. Yegros-Yegros 2012. "Science and Technology Studies: Exploring the Knowledge Base". *Research Policy* 41, 1182–1204.

Martin, B. and E. Richards 2001. "Scientific Knowledge, Controversy, and Public Decision-Making". In: S. Jasanoff, G. Markle, J. Petersen and T. Pinch (eds.), *Handbook of Science and Technology Studies*. Thousand Oaks: Sage Publications, 506–526.

Matsumoto, N. 2006. "Analysis of Policy Processes to Introduce Bus Rapid Transit Systems in Asian Cities from the Perspective of Lesson-Drawing. Cases of Jakarta, Seoul, Beijing". Institute of Global Environmental Strategies, Japan.

Mbembe, A. and S. Nuttall 2008. "Afropolis. Introduction". In: ibid. (eds.), *Johannesburg. The Elusive Metropolis*. Durham: Duke University Press, 1–33.

Mbembe, A. and S. Nuttall 2004. "Writing the World from an African Metropolis". *Public Culture* 16:3, 347–372.

McCann, E. and K. Ward 2013. "A Multi-Disciplinary Approach to Policy Transfer Research. Geographies, Assemblages, Mobilities and Mutations". *Policy Studies* 34:1, 2–18.

McCann, E. and K. Ward 2010. "Relationality/Territoriality. Toward a Conceptualization of Cities in the World". *Geoforum* 41, 175–184.

McCann, E. 2011. "Veritable Inventions. Cities, Policies and Assemblage". *Area* 43:2, 143–147.

McCann, E. 2008. "Expertise, Truth and Urban Policy Mobilities. Global Circuits of Knowledge in the Development of Vancouver, Canada's 'Four Pillar' Drug Strategy". *Environment and Planning A* 40, 885–904.

McCormick, D., H. Schalekamp and D. Mfinanga 2016. "The Nature of Paratransit Operations". In: R. Behrens, D. McCormick and D. Mfinanga (eds.), *Paratransit in African Cities. Operations, Regulation and Reform*. Oxon, New York: Routledge, 59–78.

McFarlane, C. and B. Anderson 2011. "Thinking with Assemblage". *Area* 43:2, 162–164.

McFarlane, C. 2011a. "Assemblage and Critical Urban Praxis. Part One". *City* 15:2, 204–224.

McFarlane, C. 2011b. "The City as Assemblage. Dwelling and Urban Space". *Environment and Planning D* 29, 649–671.

Mejía-Dugand, S., O. Hjelm, L. Baas and R. Ríos. 2012. "Lessons from the Spread of Bus Rapid Transit in Latin America". *Journal of Cleaner Production* 50, 82–90.

Melbye, D., L. Møller-Jensen, M. Hoppe Andreasen, J. Kiduanga and A. Gravsholt Busch 2015. "Accessibility, Congestion and Travel Delays in Dar es Salaam. A Time-Distance Perspective". *Habitat International* 46, 178–186.

Mirondo, R. 2020. "Tanzania Government Seeks New Rapid Bus Operator" Accessed at *The Citizen Tanzania* (https://www.thecitizen.co.tz/news/1840340-5461458-af0wq7/index.html). Published 19.02.2020, retrieved 17.04.2020.

Mitchell, T. 2002. *Rule of Experts. Egypt, Techno-Politics, Modernity*. Berkeley: University of California Press.

Mkalawa, C. and P. Haixiao. 2014. "Dar es Salaam City Temporal Growth and its Influence in Transportation". *Urban, Planning, and Transport Research* 2:1, 423–446.

Mngodo, E. 2016. "Living a Fast Life in a Slow City Inside the Blue Bus". Accessed at *The Citizen Tanzania* (http://www.thecitizen.co.tz/magazine/soundliving/Living-a-fast-life-in-a-slow-city-inside-the-blue-bus/1843780-3256674-32b6cy/index.html). Published 19.06.2016, retrieved 17.04.2020.

Muñoz, J. and Paget-Seekins, L. 2016. "The Promise of BRT". In: ibid. (eds.), *Restructuring Public Transport through Bus Rapid Transit. An International and Interdisciplinary Perspective*. Bristol: Policy Press, 1–13.

Myers, G. 2011. "Why Africa's Cities Matter". *African Geographical Review* 30:1, 101–106.

NBS 2013. "Tanzania in Figures 2012". Dar es Salaam. (*unpublished*).

Nkurunziza, A., M. Zuidgeest, M. Brussel, and F. Van den Bosch 2012. "Spatial Variation of Transit Service Quality Preferences in Dar-es-Salaam". *Journal of Transport Geography* 24, 12–21.

Ong, A. and S. Collier (eds.) 2005. *Global Assemblages. Technology, Politics, and Ethics as Anthropological Problems*. Malden, MA: Blackwell

Parnell, S. and E. Pieterse 2015. "Translational Global Practices. Rethinking Methods and Models of African Urban Research". *International Journal of Urban and Regional Research*, 1–11.

Parnell, S. and E. Pieterse 2014. "Africa's Urban Revolution in Context". In: ibid. (eds.), *Africa's Urban Revolution*. London, New York: Zed Books, 1–17.

Peck, J. and N. Theodore 2015. *Fast Policy. Experimental Statecraft at the Thresholds of Neoliberalism*. Minneapolis, London: University of Minnesota Press.

Pineda, A. 2010. "How Do We Co-Produce Urban Transport Systems and the City? The Case of Transmilenio and Bogotá". In: I. Farías and T. Bender (eds.), *Urban Assemblages. How Actor-Network Theory Changes Urban Studies*. Oxon, New York: Routledge, 123–138.

Pirie, G. 2014. "Transport Pressures in Urban Africa. Practices, Politics, Perspectives". In: S. Parnell and E. Pieterse (eds.), *Africa's Urban Revolution*. London, New York: Zed Books, 133–147.

Prince, R. 2017. "Local or Global Policy? Thinking about Policy Mobility with Assemblage and Topology". *Area* 49:3, 335–341.

Prince, R. 2012. "Policy Transfer, Consultants and the Geographies of Governance". *Progress in Human Geography* 36:2, 188–203.

Rabinow, P. 2005. "Midst Anthropology's Problems". In: A. Ong and S. Collier (eds.), *Global Assemblages. Technology, Politics, and Ethics as Anthropological Problems*. Malden, MA: Blackwell, 41–53.

Rasmussen, J. 2012. "Inside the System, Outside the Law. Operating the Matatu Sector in Nairobi". *Urban Forum* 23, 415–432.

Rizzo, M. 2017. *Taken for a Ride. Grounding Neoliberalism, Precarious Labour and Public Transport in an African Metropolis*. Oxford: Oxford University Press.

Rizzo, M. 2014. "The Political Economy of an Urban Megaproject. The Bus Rapid Transport Project in Tanzania". *African Affairs* 114:455, 1–22.

Rizzo, M. 2011. "'Life is War'. Informal Transport Workers and Neoliberalism in Tanzania 1998–2009". *Development and Change* 42:5, 1179–1205.

Robinson, J. 2013. "'Arriving at' Urban Policies / the Urban: Traces of Elsewhere in Making City Futures". In: O. Söderström, S. Randeria, D. Ruedin, G. D'Amato and F. Panese (eds.), *Critical Mobilities*. New York: Routledge, 1–28.

Robinson, J. 2011. "Inaugural at University College London/UCL Department of Geography". Accessed at *University College London* (https://www.ucl.ac.uk/stream/media/swatch?v=ac9c64d14960). Published 17.02.2015, retrieved 17.04.2020.

Robinson, J. 2002. "Global and World Cities. A View from off the Map". *International Journal of Urban and Regional Research* 26:3, 531–554.

Rottenburg, R. 2009 (2001). *Far-Fetched Facts. A Parable of Development Aid*. Cambridge, MA: MIT Press.

Roy, A. 2012. "Ethnographic Circulations. Space – Time Relations in the Worlds of Poverty Management". *Environment and Planning A* 44:1, 31–41.

Salon, D. and E. Aligula 2012. "Urban Travel in Nairobi, Kenya. Analysis, Insights, and Opportunities". *Journal of Transport Geography* 22, 65–76.

Salter, M. 2013. "To Make Move and Let Stop. Mobility and the Assemblage of Circulation". *Mobilities* 8:1, 7–19.

Schwanen, T. 2017. "Geographies of Transport III. New Spatialities of Knowledge Production?" *Progress in Human Geography* 42:3, 1–10.

Schwanen, T. 2016. "Geographies of Transport I. Reinventing a Field?" *Progress in Human Geography* 40:1, 126–137.

Sheller, M. and J. Urry 2006. "The New Mobilities Paradigm". *Environment and Planning A* 38, 207–226.

Sohail, M., D. Maunder and D. Miles 2004. "Managing Public Transport in Developing Countries. Stakeholder Perspectives in Dar es Salaam and Faisalabad". *International Journal of Transport Management* 2, 149–160.

Temenos, C. and E. McCann 2013. "Geographies of Policy Mobilities". *Geography Compass* 7:5, 344–357.

The Citizen Tanzania 2016. "Gleaming New Buses Challenge Chaotic Old Ways in Tanzania". Accessed at *The Citizen Tanzania* (http://www.thecitizen.co.tz/News/Business/Gleaming-new-buses-challenge-chaotic-old-ways-in-Tanzania/1840414-3355910-cpkcr7/index.html). Published 24.08.2016, retrieved 17.04.2020.

UN-Habitat 2015. "Habitat III". Accessed at *UN Habitat* (http://unhabitat.org/habitat-iii/). Published 2015, retrieved 17.04.2020.

Ureta, S. 2015. *Assembling Policy. Transantiago, Human Devices, and the Dream of a World-Class Society*. Cambridge, MA: MIT Press.

Ureta, S. 2014. "The Shelter that Wasn't There. On the Politics of Co-Ordinating Multiple Urban Assemblages in Santiago, Chile". *Urban Studies* 52:2, 231–246.

Vecchio, G. 2017. "Democracy on the Move? Bogotá's Urban Transport Strategies and the Access to the City". *City, Territory and Architecture* 4:15, 1–15.

Ward, K. 2011. "Policies in Motion and in Place. The Case of Business Improvement Districts". In: E. McCann and K. Ward (eds.), *Mobile Urbanism. City and Policymaking in the Global Age*. Minneapolis: University of Minnesota Press, 71–95.

Watson, V. 2009. "Seeing from the South. Refocusing Urban Planning on the Globe's Central Urban Issues". *Urban Studies* 46:11, 2259–2275.

Wise, J. 2005. "Assemblage". In: C. Stivale (ed.), *Gilles Deleuze: Key Concepts*. Trowbridge: Cromwell Press, 77–86.

Wood, A. 2015a. "Competing for Knowledge. Leaders ad Laggards of Bus Rapid Transit in South Africa". *Urban Forum* 26, 203–221.

Wood, A. 2015b. "Multiple Temporalities of Policy Circulation. Gradual, Repetitive and Delayed Processes of BRT Adoption in South African Cities". *International Journal of Urban and Regional Research* 39:3, 1–13.

Wood, A. 2015c. "The Politics of Policy Circulation. Unpacking the Relationship Between South African and South American Cities in the Adoption of Bus Rapid Transit". *Antipode* 47:4, 1062–1079.

World Bank 2017. "Tanzania: Rapid Bus Transit System Saves Time and Money". Accessed at *YouTube* (https://www.youtube.com/watch?v=a_V29yVYqGo). Published 26.01.2017, retrieved 17.04.2020.

World Bank 2015a. "New $300m Development Financing to Improve Key Services for 2 Million People in Tanzania's Largest City". Accessed at *The World Bank* (http://www.worldbank.org/en/news/press-release/2015/03/02/new-300 m-development-financing-to-improve-key-services-for-2-million-people-in-tanzanias-largest-city). Published 02.03.2015, retrieved 17.04.2020.

World Bank 2015b. "Project Appraisal Document on a Proposed Credit in the Amount of SDR 207.10 Million (US$ 300.00 Million equivalent) to the United Republic of Tanzania for a Dar es Salaam Metropolitan Development Project". Washington, DC.

World Bank 2008. "Tanzania Receives Credit for Second Central Transport Corridor Project". Accessed at *The World Bank* (http://web.worldbank.org/WBSITE/EXTERNAL/TOPICS/EXTTRANSPORT/0,contentMDK:21781065~menuPK:337153~pagePK:64020865~piPK:149114~theSitePK:337116,00.html). Published 27.05.2008, retrieved 11.11.2018.

Wright, L. and W. Hook 2007. *Bus Rapid Transit Planning Guide*. New York: ITDP.

2 Gathering DART

This chapter provides insights into the background and the framework of my ethnography by linking methodological considerations to my own empirical research. I explain the concept of 'gathering' and why the focus on different kinds of mobilities is important for this research. I also reflect on my positionality in the postcolonial context. Hence, this chapter discusses not only the concrete methods I applied, but also broader conceptualisations of my perspective. It discusses the question, which is particularly relevant in light of the inconsistency and deviation, ambiguity and controversy around the global BRT (bus rapid transit) model and its assembling in Dar es Salaam: how can researchers widen their perspective to diverse forms of knowing, ambiguity and ambivalence, and how can they portray (in)coherence or (in)consistency?

Rather than being a straightforward process of description, ethnography can be thought of as an orientation and a way to problematise the world. Through ethnographic approaches, researchers can explore the messy and complex nature of the (social) world. Instead of searching for generalisations or predictable patterns, they can include the mess, ambivalence, elusiveness and multiplicity (see Hine 2007; O'Reilly 2009). Ethnographies do not create coherent realities, even if assembling them is an act of ordering. The ethnographic 'weaving together' of stories brings with it uncertainty, plurality and incompleteness (Law 1993: 2). Considering the empirical as a path to the conceptual and the conceptual as practical and material (Gad and Ribes 2014: 185–186), the empirical and the conceptual are not mutually exclusive. STS-inspired approaches promote 'a theoretically informed empiricism, and a commitment to experimentation in empirical research' (Barry 2013: 419). These approaches dissolve borders between theory, methodology and method, as well as those between the perspectives of the research partners and the descriptions of the researcher.

This research is founded on the insight that methods themselves are performative; they direct the outcomes of the research. The way that I look at things shapes how I investigate them, interact with them, understand and analyse them. The use of certain methods creates rather than represents realities since methods produce different objects, and hence different realities.

Generally, realities are interactive, remade, indefinite and multiple; they are contested and might be challenged through other representations or another researchers' 'hinterland' (Law 2004, 2009). I, as a researcher can be regarded as a constructor of reality who may not hide behind 'portrayals of method as mere technique' (Hine 2007: 663), which means that science and research are performative and constitutive. Assumptions prognosticate their outcomes, or, as Latour (1996: 180) puts it: 'The human sciences do not show up as the curtain falls, in order to interpret the phenomenon, they constitute the phenomenon'.

This is my personal story of Dar es Salaam Rapid Transit (DART). Technocrats are not the only ones who tell stories about BRT and how the transport model is assembled in a specific locale. I do too because I am obviously not neutral; I possess my own ideas and knowledge. During field research, I tried to stay open by not explicitly searching for the major topics I expected, in accordance with Barry's (2013) observation:

> In practice, the researcher necessarily comes into the field equipped with a range of pieces of theoretical equipment, which may need to be tried out, modified or abandoned, but never simply applied.
>
> (418)

This ethnography is limited in two terms. First, describing an ongoing process is challenging and filled with inherent constraints. The temporal focus of this book lies on the period of planning, constructing and taking into operation of Phase 1 between 2014 and 2019.[1] Conducting major field work in 2015 and 2016, I found that it was worth focusing on the gradual and prolonged process of inaugurating DART's operations. Events that occurred before 2014 remain rather vague because I can only refer to written and oral sources. This circumstance reflects in the assembling of present and absent protagonists and other (non)humans like people and artefacts in this book. I refer more to the current CEO of the DART Agency than to his predecessors, to the updated *BRT Planning Guide* than to the first edition of the guide and to the Jangwani controversies than controversies concerning Kariakoo. Second, this writing might become emotional at times. Even if I do not aim to assess whether DART is a failure or a success or to what extent BRT is the one and only transport solution for cities like Dar es Salaam, I do have developed an opinion of and a relationship to this transport system. When DART began to matter to me, I recalled the article on the Zimbabwe Bush Pump, a technical device that became important and admirable for De Laet and Mol (2000). The authors describe how they aim to exercise neither neutrality nor general normativity regarding this fluid technology. By using the notion 'love', they reveal how affection moves their analysis of the pump and allows themselves to 'be moved' by it (253). At the peak of DART's crisis in late 2015, I felt deeply concerned that the transport system would never become operational. All the more touched I was when I saw DART's buses

Figure 2.1 Riding along the corridor on a DART bus (author's photo 10/2015).

transporting passengers on the BRT corridor a few months later. From then on, I spent numerous hours using the system, getting to know it better. I enjoyed being on the bus, moving along the corridor, flowing through the city (see Figure 2.1).

Ethnographies of gathering and assembling

I conceive my 'method assemblage' as a form of 'gathering' (Latour 2004; Law 2004). The gathering approach lets me include thoughts on temporalities and spatialities, expertise and postcoloniality, challenges and constraints. Latour understands gathering as a new form of critique and a way of detecting '*how many participants* are gathered in a *thing* to make it exist and to maintain its existence' (Latour 2004: 246, emphasis in original). In contrast, Law (2004: 100) applies the term gathering as a metaphor for bundling, and as a way of talking about ramifying relations. 'To gather is to bring "to-gether". To relate. To pick (as with a bunch of flowers). To meet together. To flow together'. More specifically:

> Gathering [is] a metaphor like that of bundling in the broader definition of method assemblage. It connotes the process of bringing together, relating, picking, meeting, building up, or flowing together. It is used to find a way of talking about relations without locating

these with respect to the normative logics implied in (in)coherence or
(in)consistency.

(2004: 160)

As such, this study does not unveil certainties but rather generates ideas and
opens up new perspectives.

Law (2009) has further coined the term 'collateral realities', by which he
means incidentally and unintentionally produced realities. Researchers
need to be careful about what they (un)consciously exclude from their realities, what they make absent. Reasons for absences vary from pragmatic
decisions (e.g. the duration of fieldwork) to affective decisions (e.g. who researchers talk to). It is essential to collect all data available and only sort it
out in the final stage. Motives and intentions should emerge and disappear
independently through the traces they leave. I am not fully aware of where
the gaps in my research are. I just know that collateral realities of DART exist. I could not include the entirety and had to reduce complexity. Researchers can regard the process of reduction and exclusion as a production of
collateral realities.

The research process itself is a process of multiple assemblings because the
(already) assembled world is (re)assembled in different stages. Assembling
connotes collage, composition and gathering (Lippert 2014: 47; McFarlane
2011: 164). To gather the elements that make up the assembling of DART,
I took an iterative, multi-locale approach to research. The field was not
bounded; I needed to follow relations to and from DART between Dar
and other locations. Second, I took the typically ethnographic approach of
'frequent movement back and forth between the specific and the more general
and between the concrete and the more abstract' (Lund 2014: 231), ensuring
that I both focused on the broader scope of this research as well as maintaining a certain distance from the subject matter. My approach required
aligning my method assemblage to the social world I was studying: a global
model being assembled in a specific locale. Thus, my research approached
the process of assembling DART by adopting its dynamics and reflecting its
shape; as intermittent and multifaceted rather than (chrono)logical.

Due to the fluid circumstances I was studying, my research had to be
both mobile and flexible. My research focus changed several times, thus
changing the nature of the field. For instance, ITDP (Institute for Transportation and Development Policy) unexpectedly opened the Africa Office in
Nairobi so that, following paths of circulation, DART led me to ITDP and
hence to Nairobi. This experience revealed new aspects of BRT dissemination in Africa, such as relations within the East African region. I needed to
be especially flexible with regard to attending the launch of DART operations, one of the major events of this infrastructural process. The launch of
DART's interim service had become a political issue and served as a powerful symbolic milestone within the controversial process of territorialising
the global BRT model in Dar es Salaam. The launch date depended on the

buses' customs clearance, which in turn depended on negotiations between transport operators, bus manufacturers, money lenders and the new Tanzanian government of 2015. Desperate to attend this event, I was in a stand-by mode in Germany for over a year, waiting for research partners to tell me when this delayed event would finally take place. Hence, ethnographic research is challenging to plan, as it is unpredictable by nature (Crang and Cook 2007: viii).

Through following the travels and changes, new questions, actors and places come into view (Till 2009: 626; Von Schnitzler 2013). Just as the DART process had been constantly shifting, so my focus and plan changed too. Most importantly, the controversies I had expected – like an open protest of relocated daladala and resettled residents – did not eventuate. Instead, other controversies appeared that made me turn from daladala workers and residents to the Simon Group, UDA and UDA-RT. Some actors are thus given less attention in this piece because it turned out that the perspectives of affected citizens (most notably those affected by expropriation and resettlement of their homes) or daladala workers (i.e. drivers and conductors) have not played a significant role in the negotiation of DART's operational model. While collecting data, I have critically reflected on why the less institutionalised actors of DART do not play a central role in this research. Additionally, I consulted employees of the DART Agency and Dar es Salaam City Council who were in charge of resettlements and compensation (see EL 10/2015; ZD 03/2015), and concluded that they were in fact not active in DART's final stages of assembling and the negotiations of the operational model.[2]

Flexibly accompanying the process of BRT mobilities and mutations by various means recalls Latour's work on *Aramis*, a private rapid transit (PRT) project developed in Paris in the 1970s and 1980s. In this ethnographic story, a professor tells his student that they need to '[s]tick to the actors. [...] If they drift, we'll drift along with them' (Latour 1996: 94). These actors – or research partners – decide what is important, and researchers need to follow these decisions. However, in my research I sometimes found it necessary to question their decisions. During interviews, when I felt that someone was not talking about a certain period of DART's planning phases because they felt it was not significant, I stopped following this period. But when I felt that certain stakeholders were trying to avoid a critical issue, for instance how local operators managed to breach their operational contracts, I carefully followed up on this omission, either during the conversation or shortly afterwards.

In order to grasp DART as global assemblage deriving from a global form (see Ong and Collier 2005), it is necessary to consider the multitude of actors and spaces, materials and discourses involved. Ethnography is itself a type of assemblage, one that traces a multitude of actions and collects and tracks diverse things (Van Loon 2014: 327). More specifically, research inspired by Actor-Network Theory, which traces relations between humans and nonhumans (see Latour 1987), enjoins ethnographers to '"follow the actors", whatever form they might take, whatever they might become and wherever they might go' (Barry 2013: 426). Regarding DART, I analyse the performativity of nonhuman actors (e.g. the buses' appearance in public space), their inscribed knowledge and policies (e.g. the buses' technical specifications) and the discourses concerning them (e.g. meanings of the buses given by other actors). By moving beyond documents, I get inside (social) organisations and events. Thereby, it is crucial to investigate a variety of human actors: not only high-ranking decision makers but also mid-level technocrats, in this case transport engineers, local consultants and operations managers (Büscher and Urry 2009: 103; Larner and Laurie 2010: 225).

How did I gather the global BRT model and the assembling of DART?[3] In order to come close to my research partners as well as to knowledge production and international standards of planning, implementation and controversies of transport models in Dar es Salaam, I combined qualitative methods, such as participant observation, document analysis and in-depth interviews.[4] Mobilities researchers need not only move with policy actors, but they need to contextualise and do 'fact-checking' (Peck and Theodore 2012: 24). I also participated in DART's implementation process from afar by following the stakeholders' WhatsApp group *BRT Education Campaign* and regularly reading Tanzanian online newspapers. Additionally, I examined various technical documents and policy papers, websites and online videos, contracts and designs. Through document analysis, I investigated 'hard facts' and looked for meta narratives, asking questions such as: How are political decisions negotiated through the material? What is the relevance of technology for 'successful' BRT operations? How do materials and things become present or absent, and how do they change shape over time? By combining these ethnographic methods, I tried to 'keep track of all moves' and bore in mind that '*everything* is data' (Latour 2005: 133, emphasis in original).

Details count a lot – for my own research endeavour and for BRT consultants who assemble designs and contracts. This also means that observation does not just involve vision but all one's senses: What do I hear and smell at the bus stations? How does it feel when the bus accelerates and people are squeezed up in it? How do I feel to be treated as a supposed BRT expert at a conference? I experienced that the shape of technical details of the buses, fences and turnstiles were not as boring as one might expect (c.f. Star 1999; see also Chapter 6). I adapted my notetaking to be as appropriate and unobtrusive according to the situation. For instance, on a DART bus, I wrote notes on my phone, a vivid mixture of English, German and Kiswahili.[5] As stated above, my inductive-iterative approach enabled me to move back and forth between theory and analysis. Consequently, during my internship with ITDP in Nairobi, my focus shifted from controversies in Dar es Salaam to global persuasive practices of transport planning; or after participating in a conference panel on STS and planning, I began to pay attention more on the nonhuman actors of DART's planning process than I had previously done. I developed the conceptual frame based on the empirical data by continuously questioning and adapting the methods to what occurs in the field. The (final) application of theoretical approaches happened only after having finished the empirical research.

Access

> I call the new CEO of the DART Agency on his mobile number, which he gave to me when I unexpectedly met him the other day. I know that he is very busy this week due to the launch of DART's service operations. He answers the call and remembers who I am: 'Ah right, you are the PhD student from Germany who is doing research on DART. You are welcome to come and visit.'[6] He tells me, almost contritely, that he is occupied today but if I could come meet him tomorrow around noon. The next day, I go back to his office, where his secretary tells me that he did not put my name on his calendar, and that he will be in meetings all day. I am frustrated but I decide to wait for an hour. Maybe he comes out of his office anytime soon so that I could at least have a short chat with him. But he doesn't. Another day lost, taking the long way to the office of the DART Agency, waiting for nothing. I start scheduling in my mind: maybe I can try to meet someone else in the afternoon. As I am about to leave, he coincidentally comes out of his office. He apologises that he forgot about our appointment, and tells me about urgent unscheduled meetings with the ministries today. He takes about twenty minutes for the most pressing questions, and asks me to come back next week, once things will have calmed down a bit.
>
> (May 2016)

Despite such frustrations, I generally had very good access to my research partners. On most occasions, I experienced easy and direct access with low barriers and only a few obstacles. I contacted high-ranking technocrats and decision makers; the doors of the service provider's offices, bus depots and workshops were always open to me. But this took me some time and effort because I needed to understand and adapt to the communication logic of my interlocutors. Besides my relationships with research partners, relationships between research partners themselves affected my access. Some partners were keen to share their perspectives with me because I had talked to a stakeholder from the opposing side; others carefully backed away from me or a specific topic I tried to address when they realised with whom I had already talked. It was common that interview partners wanted to know with whom I had already talked and what they had told me. Underlying power structures are sometimes hard to determine from the outside (Cochrane 1998).

In such a contested field, negotiations of power were an integral part of my research and were reflected in my practical experiences. Certain political issues were difficult to address; when such topics arose, research partners would either gesture dismissively and say 'politics' – a sign that they could not talk about it – or they would ask for my assurance of confidentiality. Civil servants and lawyers were particularly careful with the information they gave me: many refused to be recorded, particularly in times of conflict between the Tanzanian government and local transport operators. A few individuals even preferred to meet me at neutral places, outside political arenas and out of their colleagues' sight. Research partners I talked to several times on different stages of DART's implementation process became increasingly open – I gradually obtained their confidence.

Research is uncertainty and does not necessarily reduce complexity; it rather creates controversies (Latour 1998: 208). My experience sometimes recalled that of Professor Norbert H. and his student, who were trying to understand why the PRT project *Aramis* had failed. Every time the professor became excited that they were 'onto something' and that 'things are getting a little clearer' for the first time, the student thought: 'I didn't dare point out to him that after each interview things "got a little clearer," only to get muddled again during the next one' (Latour 1996: 99). Similarly, during interviews, I often experienced a feeling of enlightenment when a research partner told me the 'real' story, or when they described someone's 'actual' reason or hidden interest; I thought that now I had finally understood the context. But as soon as I talked to the next person, these supposed facts would come into question as a new version of the real story emerged. Particularly during strategic negotiations and controversies, I often felt like I was researching for a crime novel and not a doctoral thesis. Local operators blamed government entities for the lack of transparency, whereas officials from the government and international organisations criticised local operators for breaking laws, regulations and agreements. Over time, I learned that there was no single story and no universal explanatory framework; rather, explanations depended greatly on the (contradictory) perspectives of the actors involved (see also Adichie 2009). I tried to be simultaneously sceptical and open towards all stories in order to take all actors seriously. It was hard to determine why the stories differed: whether actors merely perceived things differently, or whether they actively tried to use me to strengthen their position by enforcing their story. The different versions I heard demonstrate the controversial and conflictual nature of DART's assembling. Thus, I adapted my methodology to account for such controversies and my research partners' complex realities (Law 2004).

Postcoloniality and language

Challenges of responsible knowledge production exist especially across the 'North–South divide' (Jazeel and McFarlane 2009: 115). On the one hand, one of my central research objectives is to contribute to ongoing decolonisation of research methods, methodology and theorisation. This academic trend has established more cosmopolitan approaches in urban studies and put African cities 'on the map' (Robinson 2002). On the other hand, I find my own research endeavour problematic. Even though I examine actors who are globally circulating and representing themselves, I also speak about (subaltern) individuals and societies that are less able to represent themselves globally (see Spivak 1988). Participatory approaches of following may diminish how much I talk about or for them; however, I am still in a powerful position since I select what is (not) written and (not) published. My opinion about conducting research as a white person from the Global North has stayed ambivalent throughout the process of this work:

> 'Why do all you German guys do research in Tanzania?' someone asks me in a pub in my neighbourhood in Dar es Salaam, sharing his observation of quite a lot of German researchers he has met in Tanzania. I do not know how to respond. He hit the mark and addressed my personal doubts and scepticism about my motivations of doing research in this setting beset with power relations and asymmetries of knowledge production. Because I do not want to justify or explain my research, I just tell him that I ask myself the same question regularly.
>
> (September 2016)

This research is to a large extent theoretically embedded in scientific discourses from the Global North. Roy (2015) makes a claim against universalisation in urban theory because this is dominated by Eurocentrism and ignores historical difference. She describes Eurocentrism as an epistemological problem (see also Mbembe 2011). Eurocentric urban theory can thus not explain (in Roy's case) Calcutta, whereby the latter could contribute to the conceptualisation of a multiple urban theory, which is concerned with the relationship between place, knowledge and power.[7] I indirectly support prevailing 'traditional' allocations: most researchers come from the Global North and from former colonial powers, whereas research partners come from the Global South, including former colonies (Lossau 2012; Macamo 2010; McLees 2013). Even if the proportion of scholars from the Global South is increasing, it is still challenging to find literature from, for instance, Tanzanian scholars. This means that I have to critically contextualise and question some (Eurocentric) perspectives applied in this work – especially those which claim to be universally applicable.

The development of anthropology as a discipline and ethnography as methodology is deeply entwined with colonial history and its legacies, especially with regard to European colonisation of Sub-Saharan Africa (Gräbel 2015; Till 2009). With this history of inequality in mind, I aimed to realise a mutual exchange by giving my research partners the position of being experts, instead of me being the superior scholar (see Czarniawska 2004: 47). Nevertheless, in this postcolonial context, I was often perceived as an expert even if I did not feel like one. Especially at the beginning of my fieldwork, I felt like an inexperienced, overwhelmed PhD student who looks underdressed in her hand-washed, wrinkled clothes, full of dust and sweat from walks in the heat to the daladala station, meeting professionals in their well-fitted suits who came to work in their comfortably air-conditioned cars.

Scholars have argued that people of colour tend to think of white people as inherently capable of taking the right decisions. As Ziai (2007: 19) describes, it is as if an actor's origin and (white) skin colour equipped them with a 'natural' authority and expertise even if they lacked local knowledge. According to Fanon (1986: 83ff.), the relationship between the (former) colonised and the (former) colonisers is always unequal and hierarchical. Centuries of white oppression and supremacy have engendered black people who feel inferior and submissive towards white people. A Tanzanian friend told me that civil servants and authorities in Tanzania act on the principle that 'white skin doesn't lie' because white people pursue 'serious' research programmes. Thus, Tanzanians often treated me as an expert, probably because white people working with DART were either from development cooperation agencies and banks, or from engineering consultancies and companies. Moreover, quite a few local actors seemed to associate my German citizenship with being experienced in high-quality infrastructure, advanced technology and reliable transport systems. Hence, my various roles and interactions as a researcher involved expectations on the part of my research partners regarding my knowledge and experience.

For my ethnographic work and my attempts to decolonise my research, it was essential to communicate in the native language of my research partners.[8] As Ndimande argues: 'The language a researcher uses in the interviews is crucial because it is through language that people formulate their thoughts as they respond to questions' (2012: 215–216). Using both Kiswahili and English has greatly enriched this work for two main reasons. First, I could dissociate myself from short-term researchers who lack Kiswahili skills, collect their data and leave. Speaking Kiswahili allowed me to build up trustful relationships. I was able to discuss personal matters with policy actors from Dar es Salaam, who were surprised to hear that I used public transport and walked around after dark, and that I knew areas outside the inner city and the expatriate peninsula. I sometimes felt like I had a kind of Kiswahili bonus that gave me access to spheres that would have otherwise remained off limits. Second, speaking Kiswahili allowed me to do research without an interpreter. Direct translations between Kiswahili and English are sometimes impossible, e.g. in situations of wordplays and idioms, both of which are common in Kiswahili.

Gathering in space and time

Since the 1980s, much has been written on multi-sited, mobile and global ethnography, questioning traditional models of small-scale research and bounded field sites (Burawoy 2001; Freeman 2012; Söderström et al. 2013). Ethnography, originally designed for 'small communities' (Tsing 2005: xi), is now routinely used to explore translocal connections. In the social sciences, increasing attention has been given to new dimensions of global exchange and movement, which provided fertile soil for the mobilities turn of the 1990s (Cresswell 2006; Sheller and Urry 2006). Out of this paradigm shift, a number of methods were developed which focus on diverse forms of (im)mobility and their social, political and economic implications; in this literature, metaphors of flux and flow, connections and relationships, abound. Mobile ethnography is the participation in patterns of movement, with a focus on processes and practices. However, the field sites are still localised to some degree, since actors are not only mobile but also situated. Therefore, it is possible to grasp processes of assembling by looking at how (non)human actors territorialise, mutate and stabilise in the specific context.

In addition, ethnographic studies can be enriched by extending observations and participations over time and space (Burawoy 1998: 17–18). McCann and Ward (2012), as well as Peck and Theodore (2012) argue for a multi-temporal and multi-directional approach, moving with the research partners in order to be able to trace mobilisations of policies. I assembled different approaches for gathering BRT's circulation and DART's formation, including its heterogeneities and (dis)continuities, relations and contradictions. I had to become mobile myself and aligned my own empirical work to the research partners. I use the lens of temporalities and spatialities to explore changes of material and discursive shapes within global forms and global assemblages. For instance, the BRT corridor in Dar es Salaam changes its shape in space, from being a clearly distinguished BRT lane to a mixed-traffic lane in the city centre. The corridor also changes over time, curtailing traffic as a construction site, accelerating traffic under DART operations, instantly losing its new and shiny appearance, accumulating traces of usage and stress marks, and being permanently under maintenance.

In order to experience everyday practices of translating the global BRT model to circumstances in Dar es Salaam, participatory approaches like thick description and 'deep hanging out' (Geertz 1973, 2000; Wogan 2004) are necessary but not sufficient for this multi-sited ethnography. Following paths of circulation can be done with tracking strategies:

These techniques might be understood as practices of construction through (preplanned or opportunistic) movement and of tracing within different settings of a complex cultural phenomenon given an initial, baseline conceptual identity that turns out to be contingent and malleable as one traces it. (Marcus 1995: 106)

Roy, drawing upon Appadurai (2001), proposes an ethnography of circulations that enables a description of the 'world of collusion and collision' (Roy 2012: 33). I understand policy making as a translocal, socio-spatial and political process that implies mobilisation and mutation. For instance, policy actors like BRT technocrats always act in between scales and at different sites. Their construction of context-specific policies is based on models elsewhere, whereby the absence of a policy in one place engenders the presence of a policy model from another locale (McCann 2011). Policies can be absent or present not only in spatial, but also in temporal terms. All combinations – policies being present or absent at a certain time in a certain locale – are worth taking into consideration.

The infrastructural process

When global policies materialise into infrastructural plans and practices in a specific locale, focusing on temporality enriches the investigation. Gupta (2015) claims that 'the temporality of infrastructure is less evident than its spatiality, but it is no less important'. Concomitant with him, Bowker suggests developing new historiographical skills, which fit into the complex natures of infrastructures:

> You never complete an infrastructure in the way you complete a novel; it is always and ever in the making. [...] It is difficult to study things that do not have a singular identity at any one moment, that do not have clear life cycles.
>
> (2015)

Especially when doing research on processes, or more specifically on the temporal dimension of infrastructures, one needs to avoid a linear or chronological understanding of temporality. The infrastructural process of DART does not follow a unidimensional, linear principle. Instead, it temporarily slows down and speeds up, unexpected actors and practices appear, and permanent change seems to be the only enduring feature. Ordering (policy) assemblages is partial, incomplete, local and implicit, and might be disconcerting for the researcher (Law 2009; Ureta 2015: 164). Consequently, I use 'temporalities' in the plural to emphasise its multiple dimensions. Additionally, I adapted my research practice to the formation and ordering practices of DART, both of which have been gradual, non-linear, and characterised by gaps and loops, detours and shortcuts:

> I sit at Ubungo Plaza, a massive building in Dar es Salaam that contains a hotel, restaurants, shops and offices. It is also the temporal location of the DART Agency. I have an appointment with the director of operations and infrastructure management for an

interview. I am too early because traffic in Dar is unpredictable so that I arranged more than enough time. Sitting on a bench next to the main entrance, I use this time to cool down from the heat of the road and to revise my questions for the director. I look at construction workers renovating the entrance area, painting walls and replacing broken flagstones. One of them puts a wooden sign next to the door with the hand-written notification: WORK ON PROGRESS. I smile as I realise that this spelling, evoking a double meaning, fits perfectly to the assembling of DART and to the assembling of this research project. After this moment, these words have decorated my pin board for a long time, reminding me that work is not only *in* progress, but that DART's stakeholders and I need to work *on* our projects' progress.

(September 2015)

Following the constantly changing shapes of DART over time in different modes allowed me to gather DART precisely and in more detail. However, every time I visited Dar es Salaam, I felt that so many important things were happening. This might partially be because I timed my field trips to coincide with events like the arrival of the buses and the launch of operations; but maybe I would have experienced similar feelings of importance regardless of when I visited the city. During my visits, I experienced a sense of time compression: the weeks in Dar expanded in my perception, whereas the weeks before and after contracted. This limitation is inevitable and another reason why this ethnography is my personal story of DART.

Timing matters in politics, and events sometimes happen unpredictably. Recent policy mobilities literature stresses the importance of tracing the process of policy circulation and to move back in time, because the entire process gives meaning to policy practices (Cochrane and Ward 2012; McCann and Ward 2012). Latour (1996: 19ff.) also undertakes a tracing-back approach to explore the local history of *Aramis*, and to search for previous critical events and issues by which his interviewees define time frames. Methodological challenges remain though. One deficiency of policy mobilities research is the often-applied retrospective angle, when scholars trace policies back to their origin. But once the process of policy circulation is completed, many steps, thoughts and controversies are no longer visible. Especially in spheres of planning and persuasion, it is worth looking beyond the knowledge that has been successfully circulated to modified or rejected policies, controversies in the planning process, interrupted flows, and so-called failures.

In my own research it proved difficult to gather from the past. When I asked about decisions from years earlier, interviewees either gave vague answers or repeated the same story over and over. In particular, the story of DART's origins has a pervasive narrative, repeated not only by research partners but also on the ITDP website (ITDP Africa 2020). According to this narrative, in 2002, ITDP staff came to Dar es Salaam and offered the City Council the chance to be part of a pilot project on BRT. The City Council was willing to participate, so funding was allocated and the initial design drawn up. In 2012, construction work for Phase 1 started. Initially, I found it hard to account for this time gap of ten years between the initial idea and the start of construction work. In time I came to realise that conceptual designs and business plans from 2007 needed to be updated and modified, study trips had to be conducted, and resettlements and compensations had proved to be more difficult to realise than expected (c.f. DCC 2007; Rizzo 2014). Unsurprisingly, however, interviewees prefer to talk about the smooth, easy aspects of the DART story. At times, they mentioned past controversies but generally left them out of the narrative, as if they were no longer significant. Another explanation for this fragmentary storytelling might be that a number of central positions have changed: some planners and managers I interviewed had not been in office at the early stages of DART.

It is important to identify not only the master narratives, but also parallel and marginalised narratives (Star 1999: 384–385). Hence, I sought to discover more about the early phase of BRT persuasion and decision making. I followed ITDP's practices of mobilising BRT elsewhere, including the initial phases of collecting ideas on transport improvements in real-time. At the time of my internship with ITDP, the NGO was promoting BRT in Nairobi and other African cities. This experience in Nairobi let me infer how complex, uneasy and twisted these early stages of BRT adoption might have been in Dar es Salaam too. Additionally, I aimed to question the dominant narratives (e.g.: 'BRT helps to solve Dar es Salaam's transport problems'; 'BRT is a good solution to stop congestion') about the City Council of Dar es Salaam's decision to adopt BRT (EL 10/2015). Moreover, I critically studied the ostensibly strong and clear arguments for BRT over light rail transit (LRT), and the one-sided advantages of a service provider model with competing bus operators – instead of promoting a business model for BRT with a single, local service provider (see Chapter 5).

I attempted to gather all stages of BRT circulation and implementation in the making, though this work does not raise any claim to completeness. The stages I detected are not fixed but rather blend into each other and might vary in order. They occur in every city differently regarding lengths, topics, etc. I followed most stages in Dar es Salaam, and some in Nairobi and from afar. Comprehensively, they are:

1   Initial idea of and decision for BRT
2   Searching for funding
3   Preparing initial design and operational model
4   Preparing on site (resettlements and compensations)
5   Constructing physical infrastructure
6   Training drivers and mechanics
7   Operational trial phase and system launch

8   Adapting and stabilising operations
9   Disseminating experiences to elsewhere

The two years of my main field research coincided with the crucial years of DART's implementation process. Several policy makers told me that in BRT projects, the difficult (and therefore most interesting) phase is the inception and implementation of BRT operations, i.e. when models and plans are put into practice. This advice convinced me to remain flexible and arrange my fieldtrips accordingly. This timing enabled me to gather the present absences (see Callon and Law 2004; Meyer 2012) of buses on the corridor and of the number plates of the buses, which directed me to the controversy of breached contracts and enforced participation of the local transport elite (see Chapter 6). I also came to understand that DART's operations – and not earlier stages of construction – were the subject of most controversies, negotiations and changes in Dar es Salaam.

Double mobility approach

This research is concerned with mobility in two main senses: BRT is a global and mobile phenomenon, and DART is a system of mobility. It was therefore necessary for me to become mobile in a double sense in order to adapt my method assemblage to the characteristics of both, DART and BRT. In Dar es Salaam, I was mobile in a sense that I frequently used DART; while studying the global BRT model, I was mobile in the sense that I participated in transport conferences and workshops in East Africa, and I joined ITDP in the Africa Office in Nairobi, which is closely connected with the head office in the US. In addition, I followed other BRT systems from afar, in their appearances of best practices in discourses and documents. I needed to simultaneously consider global movements and sites of adaptation. My methodological approach hence entails becoming mobile myself in order to come closer to my research partners, and to behave in similar ways to them. Several researchers have developed mobile method assemblages (Evans and Jones 2011: 849). For example, Verne emphasises the need to move through and between the multiple sites, in order to grasp mobility itself by 'actual moving with, accompanying and joining in mobility' (2012: 51). The researcher's movements happen in different ways, i.e. not only spatially, but also across institutional settings and disciplines: 'The researcher must move towards the subject or topic of research, whether a people located somewhere or a database stored in a physical or virtual repository' (D'Andrea et al. 2011: 153).

   International policy actors involved in circulating policies play a crucial role in policy mobilities research. Commonly, scholars studying policy mobilities only partially follow their research partners. They generally decide either to participate at the local or the global level, and many works therefore lack a deep participation in processes of policy mobilisation and mutation (see Cook and Ward 2012; Wood 2014a). By contrast, my approach of gathering not only addresses abstract global circulations but also provides insights into a concrete process of implementation, with all its tensions, adaptations and daily practices (see Prince 2012: 197). A number of scholars advocate studying middling technocrats instead of the subaltern; Roy calls this a shift from studying down to studying up (2012: 37; see also Larner and Laurie 2010). She characterises this approach as a 'mode of defamiliarization' – rendering the familiar strange, instead of rendering the strange familiar: 'I had to make the familiar "strange", paying attention to the forms of power and privilege that I often take for granted in my everyday life' (Roy 2010: 38). Similarly, De Laet (2000: 167) invites ethnographers to render their research objects and partners strange by not taking them for granted. This means that scholars can find out more by talking to and observing actors on the same level of expertise and power.

This approach matters especially in contexts of policy circulations in globally hegemonic spheres. The world of development expertise is shaped by knowledge produced by transnational organisations, which are dominated by actors who are often socialised and educated in the Global North. Nonetheless, I found it rewarding to follow not only global policy actors like European engineers and US-educated World Bank employees, but also Tanzanian individuals and institutions, and their (trans)local relations. However, this approach proved difficult because I found that Tanzanians working primarily in national or local contexts were much harder to access from a distance than research partners working in international settings like the World Bank and ITDP. This discrepancy might affect the representation of the different groups in this work. For instance, when investigating the dispute about a tender for a second service provider, I was able to gain firsthand perspectives from a World Bank advisor and the CEO of the DART Agency, but not from UDA-RT managers. Nevertheless, when I was not physically present in Dar es Salaam, I always compared perspectives from global consultants with what I read in the WhatsApp group and Tanzanian newspapers.

A common technique in studying the mobility of global policies is to follow both the policies and their mobilisers in space and time. To this end, some scholars suggest following people, policies and places (McCann and Ward 2012) while others advocate a focus on people, materials and meetings (Wood 2016). Sites and situations of policymaking are various; they can be cafés, conference rooms, online forums or offices. Wood recommends participating in ordinary practices that form the assemblages, e.g. by engaging with practitioners and their material solutions. Her experimentation with 'a Latourian approach to materiality' (393) might recall other STS works, such as Law's on the Portuguese vessel and long-distance control (1986) in which he develops three classes of emissaries: documents, devices and drilled people. An exploratory method of tracing policies – which tend to be abstract, nonlinear and unrepeatable – includes following processes, practices, discourses, technologies or networks, and thereby connecting sites, scales and subjects (see Cochrane and Ward 2012; Peck and Theodore 2010, 2012). The most obvious way to follow policies is by following people: individuals and communities, or members of organisations and institutions, through which they exert influence. Most people I dealt with were policy actors – either governmental or non-governmental, including technocrats and consultants, transport planners and engineers. Many had an international background, often through their involvement in development programmes. A second group comprised locally embedded actors like Tanzanian entrepreneurs, teachers, drivers and technicians. I found it particularly fruitful to follow mid-level actors because they could provide information about links between decision making and the executive level. In addition to the people that build global policy networks of circulation, things also mobilise knowledge and policies. A classic 'follow the thing' approach (Cook et al. 2004; Cook and Harrison 2007) could have brought me to interesting actors and places. But even if I did not follow in detail the value chain of the gearbox of a BRT bus, the production process of the steel of the stations or of the concrete of the corridor, I took 'follow the thing' as an inspiration to consider the assembling of material parts of DART. In particular, investigating the temporally differing scripts (c.f. Akrich and Latour 1992; Ureta 2015) of the buses provided deep insights into a process' frictions of global circulation and adaptation.

Since mobile ethnographies take unexpected trajectories (Marcus 1995: 96), I not only followed (non)human actors but also considered them in their different shapes and related them to one another. Also, within the DART project, actors have changed and become multiple. For instance, the contract on service provision changed during power negotiations and the BRT corridor's shape transformed over time (see Chapters 5 and 6). DART allowed me to consider both policies and narratives arriving at and departing from a specific site (see Robinson 2013). Gathering the actual BRT system in Dar es Salaam and the more abstract BRT model implies considering the global as produced in the local: 'The "local" no longer opposes but constitutes the global' (Burawoy 2001: 158). This means that global BRT would not exist without its materialisations, whether in Bogotá, Jakarta or Dar es Salaam. Only the collection of dense ethnographic material facilitates addressing policy proliferation, circulation, mutation and implementation.

Riding along

The term 'riding along' derives from the method of 'going along', which is a hybrid between participant observation and interviewing. Walk alongs (on foot) and ride alongs (on wheels) aim to access transcendent and reflexive aspects of lived experience in situ in order to find out how individuals engage with their social and physical environments (Kusenbach 2003; see also Bissell 2010).[9] The rides with DART were very inspirational. During rides, I deeply understood how the global BRT model assembles in Dar es Salaam, and how DART changes over space and time. When I had time to kill, I just took a DART bus and let my thoughts flow, together with the flowing BRT traffic. In this way of mobility and flow, my mind was awake and mobile, connecting and developing thoughts. I was anonymous inside the system, interacting with it, and at the same time inside my mental research bubble. This form of being locally mobile is a central part of my double mobility approach and an adaptation to the research subject. DART is moving within the city, and I moved with it. In order to deeply understand the system, I became a regular passenger, travelling from Gerezani to Ubungo, from Morocco to Kivukoni, back and forth. Bringing the subject of my research and method assemblage closer together means to be simultaneously on the move (Wood 2016: 393, see also Cresswell 2006), which also means experiencing the research bodily: squeezing into the hot, loud, fast-moving bus with other passengers, feeling the joints between the concrete slabs under me, and watching the traffic jam on the mixed-traffic lanes.

Being a passenger with a new comprehension of how space and time can feel on Morogoro Road – a passenger who no longer just adapts their daily travel to peak hours and congestion but now also to DART's routes:

> For the first time, I am riding along the DART corridor, together with university students and employees of the company that is constructing DART's corridor and stations. We are sitting in a regular city bus because BRT-specific buses have not been manufactured yet. On our right and left I see vehicles stuck in traffic. I get a first feeling for the flow of DART even if we regularly need to decelerate or stop because motorbikes and carrier bicycles block our way.
>
> (March 2015)

> After only a couple of minutes into my first BRT training ride, an accident occurs. A car and a motorbike, both illegally crossing the BRT corridor, have crashed and now block the lane. For training purposes, we have a police officer on board. He solves the situation after another couple of minutes. I take a look at the inside and outside of the bus. While the exterior of the bus looks quite worn, scarred by scratches and dents after less than a month of driver trainings, the interior looks like a bus that has not yet been in use. Displays tell in odd English: 'Passengers, please don't smoking compartment'. Seats are still wrapped in plastic foil. A persistently beeping sound is getting on our nerves. I am told that it will only be repaired next week when the Chinese technicians arrive in Dar es Salaam. Future DART drivers have their second day in the driver's seat. Most of them behave quite diffidently, the steering wheel shaking slightly and jiggling between their hands. The air

is sticky and hot. People along the corridor take pictures of the prototype bus in its shiny sky-blue. Pedestrians, walking on the BRT lanes, are startled when they suddenly realise that the bus is right behind them, silently approaching because of its rear engine.

(September 2015)

The bus is full; full of journalists, World Bank staff, UDA-RT management and the regional commissioner of Dar es Salaam. The rest of the bus is filled up with bystanders who want to witness this official first trip of DART's Phase 1. I am one of them, talking to an expert in automated fare collection systems from Bogotá who is surprised about the multitude of inspirations that DART got from *Transmilenio*. The next day, a friend of mine posts on *facebook*: 'Free rides for everyone!' Buses are heavily overloaded. Some passengers ride up and down the corridor, just for fun. What an experience! Consequentially, the CEO of the bus operator publicly declares that all passengers have to disembark at the terminus so that other passengers have a chance to embark. Passengers used to daladala practices spare no pains to get into the buses. At Kimara Terminal, I try to embark one of the buses. I finally make it inside, paying for the ride with a sore left foot because other passengers trod on it. Before setting off, the bus driver shows three single shoes to the passengers: Who lost them in the scramble? After the shoes found their way back to their owners, the driver turns on the music and starts the journey.

(May 2016)

Tanzanian pop music is playing inside the overloaded bus. The air is still sticky and hot. But luckily this time, the beeping sound only appears when doors are opening and closing. Some English-written rules and pictograms are now translated into Kiswahili. Sheets of paper explain: 'It is not permitted to stand here' (*Hairuhusiwi kusimama hapa*), and: 'These seats are for people with disabilities' (*Viti hivi ni kwa wenye ulemavu*). Neither pedestrians nor unauthorised vehicles trespass into the BRT corridor any more. Drivers do their job with more self-confidence, and passengers show less excitement than at the beginning of operations.[10]

(September 2016)

I had the role of a student too, and was supervised by the driving instructors and police officers. This experience was an important addition to the learning materials I received from theoretical lessons and the interviews that I conducted with technicians employed by the Chinese bus manufacturer. I observed things that I would never have learned in interviews: the abrupt, hard braking of the learner drivers in the first days (that did not comply with the teachers' understanding of passenger comfort), and the learner drivers' initial difficulties approaching the stations to be close enough for the passengers to (dis)embark. Moreover, I directly interacted with the drivers and other passengers, discussing experiences riding the bus, moving along the BRT corridor. When I joined the practical driver training on the road, I could observe first-hand processes of knowledge transfer and how this knowledge was applied to make a BRT bus move:

> On the first day of drivers' training, a driving instructor presents one of the buses to its future drivers, rallying around the rear of the bus (see Figure 2.2). The instructor ceremonially opens the hatchback and tells the future drivers that being a BRT driver is much more than just driving the bus. At the beginning of each shift, drivers need to inspect the bus from the outside and inside: hydraulic oil level, tyre pressure, battery status. He shows them the engine and warns them that their future driving experience will be different. Most of them drove lorries or minibuses before. First, since the engine has a lot of horsepower and is located at the back of the vehicle, drivers do not hear it loudly and might misinterpret their speed. They might be faster than they expect,

Figure 2.2 A driving teacher presents the rear engine to future DART drivers (author's photo, 10/2015).

and thus always need to check the speedometer (which is actually functioning, in contrast to most speedometers in daladala). Secondly, drivers need to keep in mind that Dar es Salaam's residents are not used to this kind of silent engine yet and might not hear the bus coming. Drivers should slowly approach the stations and drive extra carefully in the city centre where DART runs in mixed traffic with pedestrians and bicycles.

(October 2015)

Hanging out at circulatory sites and distant participations

In addition to participating in drivers' training, I went to other sites of BRT knowledge production and circulation, most importantly to the ITDP Africa Office, where I worked for almost three months.[11] I also attended several East African transport conferences where transport and policy actors build networks and share experiences:

> At the regional summit in Dar es Salaam, the CEO of the DART Agency and his colleagues choose to sit next to me. The CEO takes a picture of him and me, with the comment: 'That's what you do at conferences: Taking pictures with important people.' Later that day, he sends me the picture on WhatsApp.
>
> (September 2016)

Conferences are key sites for producing and exchanging knowledge, information, experience and expertise. Cook and Ward describe them as 'trans-urban policy pipelines', by which they mean 'the formation of relationships over distance as a means of comparing, educating and learning about experiences of other cities' (2012: 139). There is a reciprocity between conferences and their 'landscapes', which may well produce (material) transformations (Craggs and Mahony 2014).

Participating virtually is becoming increasingly important in ethnographic research. It opens up new configurations of space, which Beaulieu calls 'from co-location to co-presence' (2010). The WhatsApp group *BRT Education Campaign* was created in May 2015 for DART's stakeholders. It has more than 150 members, including the CEOs of the DART Agency and UDA-RT, as well as transport consultants from the World Bank and ITDP. Through that group, I was able to follow very closely special events like anniversaries and international visits, as well as proceedings and challenges of the implementation process such as overcrowding, network breakdowns or inappropriate use of the dedicated BRT lane. Particularly for the periods between my trips to Dar es Salaam and after my last visit, this group was very helpful and informative because members not only discussed current issues but also posted many photos and videos. I was generally a rather passive member of the group, but I sometimes also contributed to the discussions.

Figure 2.3 The first slide of ITDP's webinar on BRT Business Planning shows DART buses (ITDP 2018).

Being a member in this exclusive group helped me to contact other relevant stakeholders. In addition to this chat group, I also participated in ITDP's new tool of disseminating knowledge and ideals: webinars. The eight that I attended dealt with topics such as DART as a practical example of BRT implementation, the update of the *BRT Standard* in 2016 and the new edition of the *BRT Planning Guide* (see Figure 2.3).

Expert positionalities

Pritchard and Vines emphasise the need to reflect on the effects of one's own positionality: 'In ethnography, there is a requirement for the researcher to carefully examine how the researcher's social positionality and broader power relations impact upon the research process' (2013: 3). At different levels, researchers engage with and within powerful institutions of knowledge production as they enter the work spheres of their research partners. Researchers need to study the lived experience of practices of work in order to find out about the nature of the actual work in all different modes, i.e. the situated, temporal, creative, interpretive, moral and committed nature (Nicolini 2009: 134–135). In his discussion of 'experts on experts', Boyer stresses that when conducting research on other experts, we should not categorise them as utterly different from ourselves. Rather, we should deal with informants as partners or counterparts, since 'they differ from us in many ways but they also share broadly the same world of representation with us' (2008: 1). Likewise, Holmes and Marcus (2005) recommend treating research partners as epistemic partners – that is, both as subjects and intellectual partners. An anthropology of policy, as Schwelger and Powell argue,

> [...] transgresses the comfortable distinction between the subject and object of ethnographic study because it involves institutional actors who are either experts themselves or keenly attuned to the institutional structures that validate anthropological expertise.
>
> (2008: 1)

Mobility studies scholars tend to categorise actors as being either local, intermediary or global (see for instance Wood 2014b: 1242, who differentiates policy mobilisers, intermediaries and local pioneers). They often focus either on a given policy's departure or arrival city without linking back to the global policy model and the broader framework of this circulatory process. Despite their heterogeneity, BRT policy actors share the same core values regarding efficiency, affordability, comfort and safety of urban transportation. These values reflect assumptions dominated by their experiences collected in cities located in the Global North, whose realities are not much shaped by minibus systems. This 'tightly knit web' of global BRT actors (Rizzo 2017) recognises and validates itself, and its personal relationships stabilise the network. For instance, the senior urban specialist at World Bank Tanzania organised study trips on transit-oriented development (TOD) and BRT to the countries where he previously used to work, and invited an old friend from the US to conduct a survey on DART. Another former World Bank official helped Tanzanian friends obtain funding for DART, and he continues to discuss DART with a friend from university, even though both are retired and live outside Tanzania. The combination of local experience and global expertise confers them powerful positions and significantly contributes to the formation of DART.

I argue against categorising policy actors that move within and between different BRT circles according to their presumed scales of activity. A brief look at DART confirms that this policy assemblage is more complex than a blend of actors who possess either local knowledge or global experience. None of my principal research partners in Dar es Salaam would totally fit into the categories of global or local:

Herbert Meyer, the chief technical advisor of the DART Agency until 2015, was born in Switzerland, studied at the University of Dar es Salaam and used to work for the World Bank in the US. He has been continuously changing his country of residence between the three continents. He advises on greater technical and political aspects, trying to mediate between the DART Agency and the World Bank. He deeply wishes for the project to succeed because he is convinced that DART is the right answer to Dar es Salaam's traffic problems.

Rachit Padmanabhan was the operations manager of UDA-RT in 2016. Previously, he gained work experience in Ahmedabad where he managed the prestigious BRT control centre. His lack of Kiswahili and his continuous comparisons of ('unorganised, slow, unreliable') Tanzania with ('organised, fast, reliable') India made him undesirable to UDA-RT management.

The Tanzanian James Kombe has been working for the World Bank as a transport specialist for decades. He moves between international decision-making networks of the World Bank and an ordinary life in Dar es Salaam, taking DART to work every day. He is regularly invited to international transport planning summits but rarely finds time to go there. Being used to English as working language, he was one of the few Tanzanians speaking exclusively in English with me.

Collaborating with peers?

My ethnographic research became particularly collaborative while working with ITDP in Nairobi, where I had access to the source of global BRT knowledge creation and dissemination. I was interested in the way ITDP's transport experts 'use knowledge to share, direct, inform and discipline' (Prince 2012: 199). Communicating daily with the US-American director and local staff (mainly Kenyan transport planners), I learned how knowledge and ideals regarding BRT and other transport systems are mobilised. Besides working in the office on non-motorised transport and BRT projects, I attended meetings and site visits, an exhibition, a workshop and a conference. Through these, I gained insights into regional and global knowledge exchange, the enforcement of certain transport ideals and so-called best practices, and the setup of networks for future collaboration.

> At the Sustainable Urban Transport Conference in Nairobi, I talk to a transport planner from the Ethiopian Ministry of Transport. He primarily associates me with ITDP even though he knows that I am university scholar who is just collaborating with the NGO for a few months. He tells me: 'We want to implement BRT in our country. But that is not easy, we know. We need your help'. A few moments later, I talk with several participants, including the CEO of the DART Agency. When participants ask the CEO questions on DART's history, he refers to me: 'I am quite new in this business, I only started working for DART in January. You should ask her; she knows more about DART's history'. Both times, I feel a bit uncomfortable – don't they overestimate my expertise? But I am also happy to give something back, to be not always the one asking.
>
> (November 2016)

Working for ITDP, I co-edited newsletter articles and website posts on DART, and contributed to the evaluation of DART according to the *BRT Standard*. Several politicians and planners I contacted were very interested in my knowledge and opinion about DART, and some of them tried to deduce from DART's process how realistic a BRT system for Nairobi might be. (Most of them had not been to Dar es Salaam after the inauguration of DART.) Moreover, I contacted numerous transport planners and stakeholders from different East African countries and international development cooperation agencies. Over time, I became an active agent of knowledge circulation by linking different actors and sites. Planners from afar approached me for updates on DART. Especially during the first weeks of DART's operations, they were keen for me to provide them with a first assessment and pictures. Likewise, transport providers and politicians in Dar es Salaam were highly interested in global public transport trends and BRT experiences from elsewhere. Even though I handled information and opinions carefully, my position as a transmitter led to ethical and moral questions regarding my own responsibility. Having certain knowledge might bring (dis)advantages for several actors, and therefore might influence power relations.

My status in these roles had sometimes contradictory aspects: as a young woman I was seen as inferior, but as a white academic from Germany I was treated as superior. Most of my research partners were male, and I expected to be treated as a person who does not understand technical details. Surprisingly, I did not experience my gender as pertinent in my interactions; by contrast, my status as an academic played a significant role:

> I sit for an interview with the regional administrative secretary in her office as a high-ranking officer enters to pick her up for a meeting. She introduces the two of us in Kiswahili and tells him that I am a PhD student from Germany who is doing research on transportation in Dar es Salaam. He immediately responds: 'Congratulations!' (*Hongera!*)
>
> (September 2016)

Hongera is a term of great respect. However, I felt that I should be the one showing respect to the officer due to his position and age – two important social categories in Tanzania. Despite my discomfort at being accorded such respect, I benefitted also from this academic status. Some research partners even emphasised that they were happy to contribute to academic research and knowledge production.

While I was working with ITDP, everyone at the office was aware of my double role as researcher and intern. Most of the time, I was treated as a normal team member. One colleague was always joking that I should ask her the personal details I needed from her for the acknowledgements in my book because she – as my research partner – was assisting me with everything she said. In order to be transparent about my intentions, I gave a presentation to the staff about my research. This proved to be very enriching for me because we critically discussed my research questions. This double role of researchers affiliated to an organisation or a project has been referred to as being 'insider and outsider' (Wood 2016: 397) or being 'researcher and practitioner' (Bachmann 2011: 363–364). Researchers can benefit from this 'fortunate situation' since it facilitates access to (elite) informants 'by moving interaction away from a hierarchical elite/researcher relation to one of mutual information exchange' (ibid.). However, I found it not always easy to fulfil these expectations and to keep the balance between being the critical researcher and the supportive colleague. This 'dialectic relation between researcher and other experts' (Ward and Jones 1999) also evoked expectations, as for example a number of research partners wished my research was more practical – that it would offer guidance on how to successfully plan, build and operate a BRT system. In particular, staff from the World Bank and DART Agency expected my research to provide guidance on how to do things differently in Phases 2 and 3. An engineer from a European construction company told me that my transdisciplinary perspective would help the company with future BRT projects in Africa. He said that the company's engineers have finally understood that they need to cooperate with social scientists in order to successfully plan and build infrastructure projects. Moreover, he and his colleagues never see how the infrastructure is adopted, since the construction company always leaves the country before people start to use the new infrastructure.

Reflecting subjectivities and positionalities is profoundly important in any ethnographic project because both are reflected in the empirical practices and their outcomes. Taking diverse forms of knowing, ambiguity and incoherence into consideration, this chapter has shown how my ethnographic approach of 'gathering' includes temporalities and spatialities as well as postcolonial critique and multiple forms of mobility. The double mobility approach allows researchers to adapt to their research subjects, as I did to BRT and DART. Instead of trying to order the mess, this method assemblage (Latour 2004; Law 2004) discusses controversies and deviations that appear during the global circulation of the BRT model and its materialisation in Dar es Salaam.

Notes

1 At the time of writing, in May 2020, BRT consultants are spreading the DART story of success across the continent. Several cities, particularly in Eastern Africa, are showing interest in BRT, and DART is receiving a great deal of international attention as the 'African BRT'. A second bus operator has been yet to come so that the interim service provider UDA-RT is still the only bus operator. Operations lack complete scheduling, and seasonal rains let the bus service halt

regularly. Funding for three out of five future phases of DART has been secured, and construction is starting slowly.

2　As Rizzo (2014, 2017) shows, daladala workers and house/landowners played a major role in the early project delays.

3　During field trips to Dar es Salaam, I gathered different stages of DART's planning and implementing process. In February and March 2015, I witnessed the final stages of constructing the physical infrastructure, the modification of the operational model and negotiations on contracts between DART's stakeholders. In September and October 2015, I explored the operational preparations by attending the practical driving lessons over several weeks, followed the debates on the arrival of buses and subsequent negotiations on the *ISP Agreement*, which was accompanied by the presidential election campaigns. In May 2016, I travelled to Dar es Salaam to observe and participate in the launch of the interim services and the first weeks of DART operations. I attended UDA-RT's move from its sister company's premises to Jangwani Depot. In September 2016, I experienced how the interim service operations settled and how DART stabilised, as well as how the system was still under construction and how it was already under repair.

4　In total, I conducted 75 in-depth interviews; mainly during six research stays in Dar es Salaam and Nairobi between 2015 and 2016. Interviews differed between a flexible guideline-based and a rather conversational style. With key persons I had several interviews and conversations. With actors I never had the chance to meet in person because they are based in other countries, I conducted interviews via video or audio call. Within the frame of participations that resembled sometimes participatory observations and sometimes observatory participations, I had countless conversations with passengers, drivers, station attendants, driving instructors, operations managers and technicians working for DART.

5　I wrote notes as they came to mind, without making an extra step of translation. In addition to collecting notes right in the moment, I sat down every day during fieldwork, either when I had breaks between appointments or at night, and wrote down everything that came to my mind. After each field trip, I digitised all remaining analogue notes and transcribed all interviews. I processed the data with computer-assisted qualitative data analysis software and developed a coding system out of the material.

6　*Kweli, wewe mwanafunzi wa Ujerumani anayefanya PhD kuhusu DART. Karibu.*

7　See also Roy (2009) on 'new geographies of theory' and Ong and Roy (2011) on 'worlding cities'.

8　I asked my interview partners whether they preferred to speak in Kiswahili or English. On most occasions, we spoke a mixture of Kiswahili and English so that everyone felt comfortable and could express oneself accurately. People in Nairobi broadly speak English or Sheng, a hybrid between Kiswahili and English. In contrast, people in Dar es Salaam generally speak Kiswahili and people with lower educational background have limited English skills. I approached drivers, passengers and station attendants directly in Kiswahili, who then gave me names like 'Citizen of Dar es Salaam' (*Msaalamu*), 'Swahili person' (*Mswahili*) or 'Tanzanian' (*Mtanzania*).

9　Similarly, Sheller and Urry (2006) propose that mobilised ethnography can involve walking with people for deeply engaging in conversation with them, and Laurier (2004, 2008) conducts research on car travel by using the motorway as workplace. An ethnographic study of roads can reveal new perspectives on the simultaneity of global circulation of knowledge, policies and context-specific practices of (im)mobility (Dalakoglou and Harvey 2012: 463).

10　Unlike most passengers, I had already gained a general understanding of BRT from systems in Cape Town, Haifa, Johannesburg, Los Angeles and Nantes.

Moreover, I had experience of using diverse MRT systems equipped with turn-stiles at stations and stop-buttons inside vehicles.

11 Since ITDP leads the global dissemination of BRT, I focused on this organisation.

References

Adichie, C. 2009. "The Danger of a Single Story". Accessed at *TEDGlobal* (https://www.ted.com/talks/chimamanda_adichie_the_danger_of_a_single_story). Published 07/2009, retrieved 17.04.2020.

Akrich, M. and B. Latour 1992. "A Summary of a Convenient Vocabulary for the Semiotics of Human and Nonhuman Assemblies". In: W. Bijker and J. Law (eds.), *Shaping Technology/Building Society. Studies in Sociotechnical Change*. Cambridge, MA: MIT Press, 259–264.

Appadurai, A. 2001. "Deep Democracy. Urban Governmentality and the Horizon of Politics". *Environment and Urbanization* 13:2, 23–43.

Bachmann, V. 2011. "Participating and Observing. Positionality and Fieldwork Relation During Kenya's Post-Election Crisis". *Area* 43:3, 362–368.

Barry, A. 2013. "The Translation Zone. Between Actor-Network Theory and International Relations". *Millennium: Journal of International Studies* 41:3, 413–429.

Beaulieu, A. 2010. "From Co-Location to Co-Presence. Shifts in the Use of Ethnography for the Study of Knowledge". *Social Studies of Science* 40:3, 453–470.

Bissell, D. 2010. "Vibrating Materialities. Mobility—Body—Technology Relations". *Area* 42:4, 479–486.

Boyer, D. 2008. "Thinking Through the Anthropology of Experts". *Anthropology in Action* 15:2, 38–46.

Burawoy, M. 2001. "Manufacturing the Global". *Ethnography* 2:2, 147–159.

Burawoy, M. 1998. "The Extended Case Method". *Sociological Theory* 16:1, 4–33.

Büscher, M. and J. Urry 2009. "Mobile Methods and the Empirical". *European Journal of Social Theory* 12:1, 99–116.

Callon, M and J. Law 2004. "Introduction: Absence – Presence, Circulation, and Encountering in Complex Space". *Environment and Planning D* 22:1, 3–11.

Cochrane, A. and K. Ward 2012. "Researching the Geographies of Policy Mobility. Confronting the Methodological Challenges". *Environment and Planning A* 44:1, 5–12.

Cochrane, A. 1998. "Illusions of Power. Interviewing Local Elites". *Environment and Planning A* 30, 2121–2132.

Cook, I. and K. Ward 2012. "Conferences, Informational Infrastructures and Mobile Policies. The Process of Getting Sweden 'BID Ready'". *European Urban and Regional Studies* 19:2, 137–152.

Cook, I. and M. Harrison 2007. "Follow the Thing. 'West Indian Hot Pepper Space'". *Space and Culture* 10:1, 40–63.

Cook, I. et al. (M. Harrison, P. Crang, M. Thorpe, C. Blake and Cath the Red) 2004. "Follow the Thing. Papaya". *Antipode* 36:4, 642–664.

Craggs, R. and M. Mahony 2014. "The Geographies of Conference: Knowledge, Performance and Protest". *Geography Compass* 8:6, 414–430.

Crang, M. and I. Cook 2007. *Doing Ethnographies*. London, Thousand Oaks, CA: Sage.

Cresswell, T. 2006: *On the Move. Mobility in the Modern Western World*. New York, London: Routledge.

Czarniawska, B. 2004. *Narratives in Social Science Research*. London, Thousand Oaks, CA: Sage.

D'Andrea, A., L. Ciolfi and B. Gray 2011. "Methodological Challenges and Innovations in Mobilities Research". *Mobilities* 6:2, 149–160.

Dalakoglou, D. and P. Harvey 2012. "Roads and Anthropology. Ethnographic Perspectives on Space, Time and (Im)Mobility". *Mobilities* 7:4, 459–465.

DCC 2007. "Consultancy Services for the Conceptual Design of a Long Term Integrated Dar es Salaam BRT System and Detailed Design for Initial Corridor. Annex Volume 8. Impact Analysis and Mitigation. Chapter 8.5. Resettlement Action Plan. Phase 1, Part A". Draft Final Report, May 2007. (*unpublished*).

De Laet, M. 2000. "Patents, Travel, Space. Ethnographic Encounters with Objects in Transit". *Environment and Planning D* 18, 149–168.

De Laet, M. and A. Mol 2000. "The Zimbabwe Bush Pump. Mechanics of a Fluid Technology". *Social Studies of Science* 30:2, 225–263.

Evans, J. and P. Jones. 2011. "The Walking Interview. Methodology, Mobility and Place". *Applied Geography* 31, 849–858.

Fanon, F. 1986 (1952). *Black Skin, White Masks*. London: Pluto Press.

Freeman, R. 2012. "Reverb. Policy Making in Wave Form". *Environment and Planning A* 44, 13–20.

Gad, C. and D. Ribes 2014. "The Conceptual and Empirical in Science and Technology Studies". *Science, Technology & Human Value* 39:2, 183–191.

Geertz, C. 2000. *Available Light. Anthropological Reflections on Philosophical Thoughts*. Princeton, NJ: Princeton University Press.

Geertz, C. 1973. *Thick Description. Toward an Interpretive Theory of Culture*. New York: Basic Books.

Gräbel, K. 2015. *Die Erforschung der Kolonien. Expeditionen und koloniale Wissenskultur deutscher Geographen, 1884–1919*. Bielefeld: Transcript.

Gupta, A. 2015. "Suspension". In: H. Appel, N. Anand and A. Gupta (eds.), *The Infrastructure Toolbox*. Accessed at *Cultural Anthropology* (https://culanth.org/fieldsights/series/the-infrastructure-toolbox). Published 24.09.2015, retrieved 17.04.2020.

Hine, C. 2007. "Multi-Sited Ethnography as a Middle Range Methodology for Contemporary STS". *Science, Technology & Human Values* 32:6, 652–671.

Holmes, D. and G. Marcus 2005. "Cultures of Expertise and the Management of Globalization. Toward the Re-Functioning of Ethnography". In: A. Ong and S. Collier (eds.), *Global Assemblages. Technology, Politics, and Ethics as Anthropological Problems*. Malden, MA: Blackwell, 235–252.

ITDP Africa 2020. "Dar es Salaam". Accessed at *ITDP Africa* (https://africa.itdp.org/cities/dar-es-salaam/), retrieved 17.04.2020.

Jazeel, T. and C. McFarlane 2009. "The Limits of Responsibility. A Postcolonial Politics of Academic Knowledge Production". *Transactions* 25, 109–124.

Kusenbach, M. 2003. "Street Phenomenology. The Go-Along as Ethnographic Research Tool". *Ethnography* 42:3, 455–485.

Larner, W. and N. Laurie 2010. "Travelling Technocrats, Embodied Knowledges. Globalising Privatisation in Telecoms and Water". *Geoforum* 41, 218–226.

Latour, B. 2005. *Reassembling the Social. An Introduction to Actor-Network-Theory*. Oxford: Oxford University Press.

Latour, B. 2004. "Why Has the Critique Run Out of Steam? From Matters of Fact to Matters of Concern". *Critical Inquiry* 30, 225–248.

Latour, B. 1998. "From the World of Science to the World of Research"? *Science* 280:5361, 208–209.

Latour, B. 1996. *Aramis or the Love of Technology.* Cambridge, MA: Harvard University Press.

Latour, B. 1987. *Science in Action. How to Follow Scientists and Engineers through Society.* Cambridge, MA: Harvard University Press.

Laurier, E. 2008. "Driving and 'Passengering'. Notes on the Ordinary Organization of Car Travel". *Mobilities* 3:1, 1–23.

Laurier, E. 2004. "Doing Office Work on the Motorway". *Theory, Culture & Society* 21:4/5, 261–277.

Law, J. 2009: "Collateral Realities". Accessed at *Heterogeneities* (http://www.hetero geneities.net/publications/Law2009CollateralRealities.pdf). Published 29.12.2009, retrieved 17.04.2020.

Law, J. 2004. *After Method. Mess in Social Science Research.* London, New York: Routledge.

Law, J. 1993. *Organizing Modernity.* Oxford, Cambridge: Blackwell.

Law, J. 1986. "On the Methods of Long Distance Control. Vessels, Navigation and the Portuguese Route to India". In: ibid. (ed.), *Power, Action and Belief: A New Sociology of Knowledge?* London: Routledge, 234–263.

Lippert, I. 2014. "Studying Reconfigurations of Discourse. Tracing the Stability and Materiality of 'Sustainability/Carbon'". *Beltz Juventa*, 32–54.

Lossau, J. 2012. "Postkoloniale Impulse für die deutschsprachige Geographische Entwicklungsforschung". *Geographica Helvetica* 67, 125–132.

Lund, C. 2014". Of What Is This a Case? Analytical Movements in Qualitative Social Science Research". *Human Organization*, 73:3, 224–234.

Macamo, E. 2010. "Entwicklungsforschung und Praxis – Kritische Anmerkungen aus der Sicht eines Beforschten". *Geographische Rundschau* 62:10, 52–57.

Marcus, G. 1995. "Ethnography in/of the World System. The Emergence of Multi-Sited Ethnography". *Annual Review of Anthropology* 24:1, 95–117.

Mbembe, A. 2011. *On the Postcolony.* Berkeley: University of California Press.

McCann, E. and K. Ward 2012. "Assembling Urbanism. Following Policies and 'Studying Through' the Sites and Situations of Policy Making". *Environment and Planning A* 44:1, 42–51.

McCann, E. 2011. "Veritable Inventions. Cities, Policies and Assemblage". *Area* 43:2, 143–147.

McFarlane, C. 2011. *Learning the City. Knowledge and Translocal Assemblage.* Hoboken, NJ: Wiley-Blackwell.

McLees, L. 2013. "A Postcolonial Approach to Urban Studies: Interviews, Mental Maps, and Photo Voices on the Urban Farms of Dar es Salaam, Tanzania". *The Professional Geographer* 65: 2, 283–295.

Meyer, M. 2012. "Placing and Tracing Absence: A Material Culture of the Immaterial". *Journal of Material Culture* 17:1, 103–110.

Ndimande, B. 2012. "Decolonizing Research in Postapartheid South Africa. The Politics of Methodology". *Qualitative Inquiry* 182:3, 215–226.

Nicolini, D. 2009. "Zooming in and Zooming Out. A Package of Method and Theory to Study Work Practices". In: S. Ybema, D. Yanow, H. Wels and F. Kamsteeg (eds.), *Organizational Ethnography. Studying the Complexities of Everyday Life.* London, Thousand Oaks, CA: Sage, 120–138.

O'Reilly, K. 2009. *Key Concepts in Ethnography.* London, Thousand Oaks, CA: Sage.

Ong, A. and A. Roy 2011. *Worlding Cities. Asian Experiments and the Art of Being Global.* Chichester: Wiley-Blackwell.

Ong, A. and S. Collier (eds.) 2005. *Global Assemblages. Technology, Politics, and Ethics as Anthropological Problems.* Malden, MA: Blackwell.

Peck, J. and N. Theodore 2012. "Follow the Policy. A Distended Case Approach". *Environment and Planning A* 44:1, 21–30.

Peck, J. and N. Theodore 2010. "Mobilizing Policy. Models, Methods, and Mutations". *Geoforum* 41, 169–174.

Prince, R. 2012. "Policy Transfer, Consultants and the Geographies of Governance". *Progress in Human Geography* 36:2, 188–203.

Pritchard, G. and J. Vines 2013. "Digital Apartheid. An Ethnography Account on Racialised HCI in Cape Town Hip-Hop". Proceedings of the SIGCHI Conference on Human Factors in Computing Systems, April 2013.

Rizzo, M. 2017. *Taken for a Ride. Grounding Neoliberalism, Precarious Labour and Public Transport in an African Metropolis.* Oxford: Oxford University Press.

Rizzo, M. 2014. "The Political Economy of an Urban Megaproject. The Bus Rapid Transport Project in Tanzania". *African Affairs* 114:455, 1–22.

Robinson, J. 2013. "'Arriving at' Urban Policies / the Urban: Traces of Elsewhere in Making City Futures". In: O. Söderström, S. Randeria, D. Ruedin, G. D'Amato and F. Panese (eds.), *Critical Mobilities.* New York: Routledge, 1–28.

Robinson, J. 2002. "Global and World Cities. A View from off the Map". *International Journal of Urban and Regional Research* 26:3, 531–554.

Roy, A. 2015. "Who's Afraid of Postcolonial Theory"? *International Journal of Urban and Regional Research* 40:1, 200–209.

Roy, A. 2012. "Ethnographic Circulations. Space – Time Relations in the Worlds of Poverty Management". *Environment and Planning A* 44:1, 31–41.

Roy, A. 2010. *Poverty Capital. Microfinance and the Making of Development.* New York, London: Routledge.

Roy, A. 2009. "The 21st Century Metropolis. New Geographies of Theory". *Regional Studies* 43:6, 819–830.

Schwelger, T. and M. Powell 2008. "Unruly Experts. Methods and Forms of Collaboration in the Anthropology of Public Policy". *Anthropology in Action* 15:2, 1–9.

Sheller, M. and J. Urry 2006. "The New Mobilities Paradigm". *Environment and Planning A* 38, 207–226.

Söderström, O., D. Ruedin, S. Randeria, G. D'Amato and F. Panese 2013. "Of Mobilities and Moorings. Critical Perspectives". In: O. Söderström, S. Randeria, D. Ruedin, G. D'Amato and F. Panese (eds.), *Critical Mobilities.* New York: Routledge, v–xxvi.

Spivak, G. 1988: "Can the Subaltern Speak"? In: Cary Nelson and Lawrence Grossberg (eds.), *Marxism and the Interpretation of Culture.* Basingstoke: Palgrave Macmillan, 271–313.

Star, S. 1999. "The Ethnography of Infrastructure". *American Behavioral Scientist* 43:3, 377–391.

Till, K. 2009. "Ethnography". In: Rob Kitchin and Nigel Thrift (eds.), *International Encyclopaedia of Human Geography* 3, 626–631.

Tsing, A. 2005. *Friction. An Ethnography of Global Connection.* Princeton, NJ, Oxford: Princeton University Press.

Ureta, S. 2015. *Assembling Policy. Transantiago, Human Devices, and the Dream of a World-Class Society*. Cambridge, MA: MIT Press.

Van Loon, J. 2014. "Reassembling Ethnographie. Bruno Latours Neugestaltung der Soziologie". In: D. Lengersdorf and M. Wieser (eds.), *Schlüsselwerke der Science & Technology Studies*. Wiesbaden: Springer VS, 319–329.

Verne, J. 2012. *Living Translocality. Space, Culture and Economy in Contemporary Swahili Trade*. Stuttgart: Franz Steiner.

Von Schnitzler, A. 2013. "Travelling Technologies. Infrastructure, Ethical Regimes and the Materiality of Politics in South Africa". *Cultural Anthropology* 28:4, 670–693.

Ward, K. and M. Jones 1999. "Researching Local Elites. Reflexivity, 'Situatedness' and Political-Temporal Contingency". *Geoforum* 30, 301–312.

Wogan, P. 2004. "Deep Hanging Out: Reflections on Fieldwork and Multisited Andean Ethnography". *Identities: Global Studies in Culture and Power* 11, 129–139.

Wood, A. 2016. "Tracing Policy Movements. Methods for Studying Learning and Policy Circulation". *Environment and Planning A* 48:2, 391–406.

Wood, A. 2014a. "Learning Through Policy Tourism. Circulating Bus Rapid Transit from South America to South Africa". *Environment and Planning A* 46, 2654–2669.

Wood, A. 2014b. "Moving Policy. Global and Local Characters Circulating Bus Rapid Transit Through South African Cities". *Urban Geography* 35:8, 1238–1254.

Ziai, A. 2007. "Rassismus und Entwicklungszusammenarbeit. Die westliche Sicht auf den Süden vom Kolonialismus bis heute". In: Berliner Entwicklungspolitischer Ratschlag (ed.), *Von Trommlern und Helfern. Beiträge zu einer nich-trassistischen entwicklungspolitischen Bildungs-und Projektarbeit*. Berlin, 12–18.

3 Fluid formations

Disseminating BRT globally requires a lot of work. Before BRT can become mobile, it must first be transformed into a 'travelling model' (Behrends et al. 2014). This model needs to be flexible enough that it can be translated into different contexts; but it also needs to be standardised to prevent just any bus system calling itself BRT. Travelling models need to be both rigid and stable as well as fluid and mutable. In order to explain why the formation of global BRT is fluid rather than rigid, I elaborate three central features of the model – mobilisation, translation and standardisation – and show that policy models are not as solid and rigid as they might seem, and as global proponents would like them to be. In order to elaborate how BRT is a travelling mode, I draw on Latour's notion of 'immutable mobiles' (1986, 1987), De Laet and Mol's conceptualisation of 'fluid technologies' (2000), and Ong and Collier's term 'global assemblages' (2005). Additionally, I consider reflections from policy mobilities literature, such as McCann and Ward's (2013) description of policy models and Freeman's metaphor of waves: 'Policy, like the wave, exists only because the elements of which it is composed are moving' (2012: 18). I show that a model's displacement and mutation are mutually dependent, and that these are the principal processes of translation. These processes lead to the fluid formation of the global BRT model, as well as the fluid formation of the specific BRT system in Dar es Salaam. I thus ask: how mutable and how standardised does the global model need to be in order to be mobile?

BRT as a travelling model

A model is the conceptualisation, standardisation and mobilisation of expertise; it concerns something transferable. The dissemination of the BRT model from Latin America to Asia and now increasingly to Africa is a challenging task. While BRT consultants and planning guides aim to reduce mutation by emphasising the importance of BRT's central characteristics, adaptation is essential; the model must be modified according to specific contexts. Furthermore, despite the power of global consultants, recipients have some influence on the model: 'Cities of the Global South are not passive receivers of international best practice, but actively engaged in shaping it to suit local conditions' (Wood 2014b: 1251). The more the new context differs from the original context in which the model was created, the more translation work is necessary. Still, for a policy to become a model it must achieve a certain degree of stabilisation. Ureta sees models and their embodied expertise as 'inscription devices' that become increasingly black-boxed (2015: 78, see also Latour and Woolgar 1986). The concept of travelling models 'proposes to emphasise *how* things and ideas move from one place to the other, which suspends the contradiction between local specificities and global abstractions' (Behrends et al. 2014: 9, emphasis in original). For the authors, models are the product of global assemblages, the articulation of a global form in a specific context. In contrast, global forms are not bound to particular sites and 'have a distinctive capacity for decontextualization and recontextualization' (Collier and Ong 2005: 11). Global forms can be technologies and policies, EU standards and ISO norms, economic and mathematical models (see also Sismondo 1999). I argue that the concept of global forms rather than global assemblages better helps to conceptualise the nature of travelling models.

Models are 'enmeshed in the transformation of institutional landscapes' because they contribute to global learning processes (Peck and Theodore 2015: xxv). A model comprises a set of rules and techniques, but it only becomes a model once it travels (Rottenburg 2009). If such a set exists only at one site, it is context-specific and its ability to be de- and reterritorialised is not necessarily given. The two notions of 'travelling' and 'model' are thus inextricably linked with one another because models promise 'extra-local salience' (Peck/Theodore 2010: 171). Material technologies put objectified ideas, which are inscribed in models, into practice. Models can thus be analytical representations of certain aspects of reality, and they can also be constructed to shape this reality for particular purposes (Behrends et al. 2014: 1–2; Rottenburg 2009).

Policy mobilities scholars no longer see policy transfer as a simple transfer from one place to another but rather as a complex, mutual process which 'resembles a rolling conversation rather than a coherent paradigm' (Peck 2011: 2). A central concern is how to understand the production and circulation of knowledge (Peck and Theodore 2015; Temenos and McCann 2013). Scholars have started to analyse local change in relation to global developments by asking how change in one place is related to developments in others (Behrends et al. 2014: 20–21; Eriksen 1991). Prince states:

> [...] Policies can be packaged into forms amenable to travel and translation on policy circuits, out of which they can be refixed elsewhere, often with different results, and occasionally resulting in a new, mutated policy approach for release back onto the circuit.
>
> (2012: 192)

Because diverse actors contribute to translating them, policies evolve differently in different settings. And because ideas lead not only to homogenisation but also to variation, translation may give rise to contingencies and unpredictable changes (Sahlin and Wedlin 2008). According to recent debates on global policies, 'mobility is an inherent characteristic of policy, not an epiphenomenon', whereby 'policy changes as it moves, and the more it moves the more it seems to change. [...] It must change in order to move, and it must move in order to exist' (Freeman 2012: 20). Therefore, policies appear in the shape of models that are inevitably both mobile and mutable. Peck describes this mutual relationship as a 'tandem' through which new geographies of policies are shaped. Moreover, he evokes the 'nascent mobility-mutations approach' that explains how the (often surprising) forms and effects of policies are transformed and hybridised by their journeys (2011: 21–22). Similarly, McCann and Ward argue that policy models are 'hybrids of hybrids', holding various characteristics from different sites and local specifications because models were 'assembled and reassembled on their travels' (2013: 14). The (re)interpreting of policies happens all the time and everywhere. Mutation occurs internally and externally. Policy models are (re)shaped by the 'ephemeral political objectives of city leaders' (Wood 2015: 1074). In turn, places, institutions and communities are reshaped through territorialising (global) policies (Cook and Ward 2012: 149). In short, policies never arrive in the same form as they were originally shaped; nevertheless, consultants strive to perpetuate a consistent story of policy models.

One concept used to understand 'the complex infrastructural conditions that allow global forms to function' (Collier and Ong 2005: 12) is Latour's 'immutable mobiles' (1986, 1987). As the name suggests, certain (technical) devices are immutable and mobile at the same time, meaning that they are contextually unbound. Drawing on Eisenstein's (1979) work on the printing press, Latour writes: 'Immutability is ensured by the process of printing many identical copies; mobility by the number of copies, the paper and the movable type' (1986: 10). Because immutable mobiles are translated into different contexts, they 'have to be able to withstand the return trip without withering away' (Latour 1986: 7). Thus, (only) through translation, can an immutable mobile 'move through space while holding its shape' (Law and Mol 2001: 619). This chapter hence discusses to what extent the global BRT model is (im)mutable.

Translation and mutation are essential processes of policy models' circulation. Inspired by classic Actor-Network Theory, I see translation as the conjunction of displacement and transformation. It is a semiotic, political and geographical process in which models, knowledge or artefacts are displaced, transformed and assembled in different shapes (Callon 1986; Latour 1994; see also Barry 2013). 'Chains of translation' describe the work of modifying, displacing and translating various and contradictory interests (Latour 1999: 311). Translation happens on the move, often in spaces in between, blurring the boundaries of scale. As Callon has shown, translation is 'never a completed accomplishment' (1986: 196). It may fail, and its result can be contested. Translation is hence a matter of power and control exercised between heterogeneous actors and materials.

Other travelling models that resemble BRT include the *Zimbabwe Bush Pump* and patents. The first case focuses on a fluid, 'changeable object that has altered over time and is under constant review' (De Laet and Mol 2000: 228). The pump's vague and moving boundaries make it adaptable, responsive and flexible. Even the core function of the pump can be questioned by undermining its 'essential' characteristics (Law 2002). The second case focuses on the agency of mobile objects. De Laet argues that patents connect different places, are able to adapt to various contexts, carry knowledge and raise expectations: 'So the patent is not a neutral medium; it *does* something' (2000: 165; emphasis in original). Echoing recent debates of policy mobilities literature, De Laet argues that patents also change on their journeys, precisely *because* they travel:

> Do the words and drawings remain the same while they travel? In principle yes, but in practice, I would say, not. For a close look at such travel suggests that it may be possible to make a compound from a patent. It may be possible to learn how an engine works from a textbook. It may be possible to repair a machine from a blueprint. But patent and compound, textbook and engine, blueprint and machine, can only be considered 'the same thing' as long as they remain in the same 'place', that is, as long as they are in a place where they are made to refer to each other.
>
> (163)

In practice, patents depend on context; they vary from site to site. In this sense they are 'boundary objects' (Star and Griesemer 1989; see also Bowker and Star 2000). Interpreted differently by different groups and heterogeneous actors, they can be simultaneously abstract and concrete, specific and general. They have different meanings in different contexts, but they can retain enough immutable content to maintain a common or universal identity. Likewise, I ask to what extent the BRT model changes its shape as it changes context and whether it has any essential characteristics.

When I asked the chief technical advisor of the DART Agency about the likelihood of Dar es Salaam Rapid Transit (DART) being economically viable, he emphasised that models might come out differently in practice: 'Our models say "yes". But models are models. [...] That's a theory. We don't know' (HM 03/2015). As this quote shows, the concept of 'model' appears in different BRT contexts: scholars and consultants refer to BRT as a 'transport model', whereby they differentiate between the physical design and the operational model; famous BRT systems are referred to as 'the Curitiba model' or 'the Bogotá model'; and certain characteristics, techniques or strategies of global BRT are built on models, e.g. the 'SP (service provider) model', 'financial model' and 'PPP (public-private partnership) model'. Models not only have direct impacts on technopolitical decision making but also (often unforeseen) effects on socio-material practice. The following section discusses the early stage of the fluid formation of global BRT, namely the model's mobilisation.

Mobilisation

Diverse tools are used to mobilise policy models such as BRT, including conferences and workshops, bi- and multilateral visits, tours and trips, reports and journals. The Institute for Transportation and Development Policy (ITDP) is the central organ of BRT mobilisation in the Global South. In the following, I discuss the main tools used to circulate the BRT model.

For global consultants, policy models serve as 'evidence', i.e. that the respective city will benefit similarly from that policy like the city of origin did. Local politicians are encouraged to assume that the BRT model will have comparable effects in their own cities like the model had in cities like Bogotá or Istanbul (GV 11/2016; Wood 2014a). Thereby, the BRT dissemination strategy presents the (alleged) success of BRT elsewhere in order to persuade local politicians of BRT (see Chapter 4). All models have their best practices, be it BRT, business improvement districts, or growth and development strategies. Best practices involve normative power, which is embedded in colonialism and globalisation (Vainer 2014: 48). Since best practices are more mobile than other models, they exert hegemonic power over distance. Generally, they move within 'power-laden circulation patterns of knowledge that are intrinsically political' (Stein et al. 2017: 37). Discursive repetition and their framing as success make them present and more famous than others (Moore 2013: 2374). Comparability is another important feature, which is imagined to be realised through simplification and abstraction. According to policy mobilisers, global learning should forestall the repetition of mistakes and failures. Most mobilisers prefer to disseminate best practices and do not believe in effective learning from failures. Policy actors perpetuate the glossy image of *Transmilenio*, disregarding the fact that other cities could also learn even more from difficulties and controversies. However, policy learning implies not only how to do things but also how not to do things. Delhi, the only official case of BRT failure worldwide (Filipe and Maćario 2013: 151; Muñoz and Gschwender 2008), could teach different lessons than Bogotá. The focus on best practices and stories of success is reflected in the scholarship on mobile policies. According to Lovell, this tendency is reproduced in the 'empirical bias towards analysis of successful policies' of policy mobility scholarship (2017: 13). Currently, early attempts are being made to consider so-called failed circulatory practices too, and to use them productively for global learning (ibid.; Stein et al. 2017).

Dominant documents

Documents are one central means of mobilising BRT; they play a crucial role in shaping the global BRT model and influencing local governments. According to Law (2011), technical documents undergo complex processes of translation before publication. Their contents derive from not only technical, but also social and political decisions. Hence, the technical is always social. The extensive digitalisation of the last decade has strongly increased the global circulation of knowledge, also regarding BRT. Technocrats pursue increased global exchange and enforced standardisation by distributing (increasingly digital) BRT documents to guide project partners. For instance, ITDP's consultants distribute handbooks at conferences and webinars with the goal of influencing transport planners' decisions. The use of normative terms like 'best practice' and 'world-class' is central in these mobilisation attempts.[1] According to Stirrat, such documents 'become templates for knowing the world and also epitomise a particular view of the world' (2000: 36). A large number of documents from development projects are characterised by a set of modernist principles, assumptions and normativities. BRT proponents endeavour to downplay difficulties of implementation processes and (operational) weaknesses of systems so that I could find hardly a word of critique in the public statements of ITDP and other proponents. However, they were very open about such difficulties in bilateral conversations.

The most well-known and widely circulated BRT document is the *BRT Planning Guide* (Wright and Hook 2007; ITDP 2017b). The *BRT Planning Guide* is unique; no other means of public transport has a similar tool for mobilising policies. An ITDP employee concluded the first webinar of a series on the updated *BRT Planning Guide*:

> The thing to remember is that BRT is a lot more than just the infrastructure. It is looking at the communications plan, it is having a business plan, it is looking at what services you going to offer, it is about the institutional set-up. It is about contracting, it is about the technology that runs it. And that's what the *BRT Planning Guide* addresses, all these things.
>
> (ITDP 2018a)

The cover of the 2007 edition shows *Transmilenio*, representing the best-practice BRT of the early 2000s. Edited by two of the most famous BRT consultants and filled with colourful graphs and pictures over approximately eight hundred pages, this edition is available online as an open access document. ITDP describes the guide as the most comprehensive resource for planning BRT systems. It is primarily aimed at planning and engineering professionals but also serves as a resource for NGOs and politicians. More fundamentally, the guide claims to be applicable to any urban context worldwide and promises to bring all elements and steps together: from initial preparation and funding, to business planning and construction, through to implementation and long-term operation. Thereby, the guide aims to represent the main feature of BRT, namely its global applicability.

Opinions differ as to how far the guide is applied in practice. In order to convince me of the guide's global mobility and widespread impact, the consultant Peter Hall told me the story of Tehran's BRT. According to this account, Tehran's city administration implemented a BRT system solely by using the *BRT Planning Guide*, a hard copy of which Iranian policy actors had somehow coincidentally received. At that time, ITDP was unaware of the city's BRT plans because no one from Tehran had contacted them. Afterwards, planners from Tehran explained that they had not seen any need to contact the consultants directly since the guide provided all the necessary information. In contrast, most policy actors in Dar es Salaam were not aware of it. However, international consultants in Nairobi and South African politicians considered the document as inspiration and guidance for planning their respective BRT systems. Some policy actors still possess a hard copy of the first edition of the guide, but most use the online version now. Wood writes that most planners she met in South Africa had a hard copy of the guide 'promiscuously placed at their desks – like the Gideons Bible in a motel room nightstand' (2016: 399). However, it turned out that few had actually read the guide: 'Its importance seemed to supersede the words themselves and the materiality of the document became as much a justification for introducing BRT as an instructive technical manual' (ibid.).

The first comprehensive update of the guide was published in 2017. The 'cleaned up' update reflects the 'tons of changes' in BRT planning of the last decade (ITDP 2018a; ER 08/2016). However, the global BRT model has not been fundamentally scrutinised since the start of global dissemination in the early 2000s. The *BRT Sourcebook* (Wright 2003), the predecessor of the *BRT Planning Guide*, describes BRT in almost identical terms as recent publications. There are several changes worth mentioning though. At over a thousand pages, the new edition is even more elaborate than the 2007 edition. Moreover, the aesthetics have changed, making BRT more social and political – almost poetic. More than 200 aphorisms accompany the rather technical descriptions on how to comprehensively plan a BRT system. Kofi Annan, Charles Darwin, Mahatma Gandhi, Steve Jobs, Immanuel Kant, Michelle Obama, Pablo Picasso, Margaret Thatcher and many more have something to say about BRT roadway design, signal control and demand analysis. These poetics were well illustrated during a webinar, when the chief programme director presented a picture from Dar es Salaam, and elaborated how the guide could mitigate risks involved in planning and investment processes:

> The picture shows a DART bus riding along the shore of the Indian Ocean. In the background, pedestrians walk around, and skyscrapers, seagulls and palms rise into the sky. The evening sky has the same colours as the bus: Blue and orange. This romantic scene is perfected with a quote of Antoine de Saint-Exupery on top of the picture: 'If you want to build a ship, don't drum up the men to fetch the wood, allocate the jobs and divide the work, but teach them the yearning for the wide open sea.'
>
> (de Saint-Exupery in ITDP 2018a)

The developments of the last ten years of BRT planning and ITDP's goal of disseminating BRT to Africa are reflected in an additional way. Since 2017, ITDP has used DART as best practice and colourful background illustration in most of its webinars and other publications. Because DART has become a crucial example for BRT in the Global South, the Tanzanian system embellishes the cover of the *BRT Planning Guide* 2017 (see Figure 3.1).

Another central feature of the 2017 edition is its blog-like format, to which a broad variety of consultants have contributed experiences and knowledge. This format claims to be open for contributions 'from outside ITDP' and functions as a 'real-time document like *Wikipedia*' (PH 08/2017). The NGO encourages a dynamic updating from the BRT community, i.e. professionals

Figure 3.1 DART at the main webpage of the *BRT Planning Guide* (ITDP 2017b).

and planners from all over the world (ITDP 2017b, 2018d). One might assume that this recent strategy of inviting 'everyone' to participate could lead to an increasing mutability of BRT. Not only the update of the *Planning Guide* but also the opening up of authorship indicate the possibility of multiple forms of BRT emerging and changing across space and time. However, rather than proliferation, consolidation happens. First, ITDP has retained its function as the supervisor of the dominant BRT document; second, the 2017 version is only available in English. The idea of a central document that is constantly under revision has not yielded parallel documents in different languages. Rather, this centralisation of the guide has engendered increased homogeneity of global BRT knowledge (the 2007 edition was published in five languages), and more control by a few central actors of the BRT community that are in fact contributing to the guide.

Spaces of encounter

The global BRT model is moreover made mobile by bringing people together, attracting visitors, and sending planners and politicians elsewhere. The policy mobilities literature emphasises the importance of conferences and study tours as common mobilisation tools, describing them as 'globalizing microspaces' (Larner and Le Heron 2002: 765). According to ITDP, 'visits to successful public transport systems, visits from successful mayors and other successful implementers [are] modes of awareness-raising [and] one of the best opportunities for knowledge transfer' (ITDP 2017b: 10). The implementation of BRT systems is often preceded by similar encounters. From Bogotá, the BRT model took the same route to Johannesburg as to Dar es Salaam, involving similar mobilisation tools. First, global BRT experts visit the respective city and raise the expectations of local policy actors by presenting shiny stories and impressive images of *Transmilenio*. Next, a local delegation of transport officials, local consultants and minibus operators travel to Bogotá to experience it for themselves. Finally, possibly after some detours and controversies, they decide on BRT and develop an infrastructural design that resembles the best practice system.

Organisations and individuals promote, organise and finance BRT study tours under the guise of philanthropy, often promoting (alleged) 'South–South policy learning' (Montero 2016: 7; Temenos and McCann 2013: 349). Study trips are also a basic tool of the World Bank, not only to demonstrate BRT best practice, but also to address topics like transit-oriented development (TOD) and land-use planning. A senior urban specialist of World Bank Tanzania described study trips as 'one of the most successful things' for connecting people and creating deep relationships between them: 'There is nothing more important than study trips. [...] You have to see these things. It's worth [more] than looking at a study' (BN 03/2015). According to him, one can teach Tanzanian decision makers significantly more on a one-week tour than could ever be achieved in Dar es Salaam. A retired World Bank consultant from Nairobi shared a similar perspective, describing study trips as a successful tool because Kenyan officials only became willing, even enthusiastic, to implement BRT after they visited *Metrobus* in Istanbul: 'Seeing is believing', he said (DN 03/2015). Opinions differ on the effectiveness of these trips though. Several scholars and consultants criticise them as unproductive because they function primarily as policy tourism (Montero 2016; Wood 2014a). For example, James Kombe, another World Bank transportation specialist elaborated:

> [...] most of the African countries, they do study tours as tourism. They don't really study anything. But if you really need to study and make sure that you don't reinvent the wheel, you don't need too many study tours. A lot of this, you can easily model them. You can sit even here and understand. You can *google*, you can go to *YouTube*, you can see how the BRT functions, how the metros function, you can do most of the stuff. So, you can complement that maybe one study tour with a specific focus. And you'll have specific questions on areas, which are not clear to you. But you don't really need to be always travelling.
>
> (JK 10/2015)

Tanzanian officials went on study trips to Bogotá in 2003 and Miami in 2009. The then CEO of the DART Agency described the Miami visit as an 'eye opener' (US Embassy Tanzania 2009). Nevertheless, the current CEO questioned the sustainability of this visit and criticised that tours were undertaken too early, before DART's stakeholders knew what they needed to consider, especially in terms of technology and operations (GV 09/2016). He suggested pursuing a more interactive form of knowledge exchange and invited his East African colleagues to Dar es Salaam: 'Deep visits instead of simple study trips!' (GV 11/2016). The idea was that they not only look at DART and hear the stories told but also work together over a longer period. Actual outcomes of study trips are hard to measure. However, they are a powerful instrument to promote policies through social bonding, experimental learning, building trust and consensus, and exchanging explicit knowledge and tacit understandings (Montero 2016: 1; Wood 2014a: 2659). They have been established as a central tool for raising awareness about BRT, and they made a lasting impression on Tanzanian government officials even if the latter did not remember crucial details for operational planning. These trips allowed Tanzanian officials to become mobile and connect to other members of the global BRT network. They also gained a level of authority through experiences they collected and the ability to refer to places elsewhere, thus legitimating their political decisions in Dar es Salaam.

International conferences and workshops provide another important way of participating in the global BRT community. Conferences are 'important arenas in and through which both the mobilizing and embedding of urban policies can occur' (Cook and Ward 2012: 137), at which policy models are developed, shaped and distributed to a multitude of participants, such as governments, development agencies and the private sector. Like study trips, conferences are selective and therefore exclusionary because they render other actors and sites invisible or unimportant. The lucky ones who have the chance to attend conferences are generally eager to expand their networks, something often explicitly encouraged by the organisers. For example, the convenors of a conference on sustainable urban transport in Nairobi stressed their goal of building a regional network. Within the framework of this regional network, UN Habitat, the German Agency for International Cooperation (GIZ) and ITDP regularly organise so-called placemaking events in participating cities to promote their sustainable transport ideals, focusing on the combination of BRT and non-motorised transport (NMT). At one such event in Nairobi in late 2016, the organisers named BRT the 'new face of public transport in Africa' (see Figure 3.2).

A third space of encounter is webinars, a relatively new tool for mobilising BRT knowledge and for global planning more generally. Planners and politicians can meet online with the global BRT crowd and famous BRT technocrats. As one of ITDP's webinar convenors explained: 'I really wanted to make sure that this work I've been working on got more out there in the public' (ER 08/2016). More than a hundred people participated in the first part

Figure 3.2 BRT exhibition in Nairobi's streets (author's photo 11/2016).

of the webinar series on the new *BRT Planning Guide*. This series started in January 2018, two months after the guide had been published online. Officially, webinars aim to ease access to the guide for policy actors interested in BRT, but they also advertise and disseminate the guide, along with its inscribed knowledge and ideals, and consequently facilitate the spread of BRT. Webinars start with an introduction by an ITDP member followed by a conference-style presentation by a BRT proponent with illustrative slides. Participants are mainly policy actors and transport planners who already have a basic understanding of BRT. During the talk, they can write comments in a chat box and add questions to a separate Question and Answer box, which the speaker answers at the end. Afterwards, ITDP employees document the webinar's content and share it with all participants. During the various BRT-related webinars, consultants addressed the six volumes of the new Planning Guide, the updates of the *BRT Standard*, and DART as best practice for 'transforming urban mobility in East Africa' (ITDP 2018g). Because ITDP pursues an open-access publishing strategy, not only digital copies of the guide but also the recorded webinars are publicly available. ITDP's values and ideals are closely reflected in webinar content, and most of the presentations are by technocrats from within ITDP's network. Hence, webinars reinforce a sense of the model's immutability rather than opening it to mutable and fluid forms of BRT. Due to the rising number of participants, ITDP considers webinars to be a successful tool (ITDP 2018c).

Strategic combinations

The third way that ITDP mobilises the BRT model is by combining two strategic tools: the annual *Mobilize* summit and the *Sustainable Transport Award* (ITDP 2017a, 2018i). The NGO presents its summit as a global highlight – a 'sustainable transport get-together' (ITDP 2018f) – that is directly connected with the *Sustainable Transport Award* (STA). With the help of an international committee of transport specialists and development actors, ITDP presents the STA to cities that have pursued innovative sustainable transportation projects. A significant number of these cities have a BRT system, described by ITDP as an 'elite group of international cities' (ITDP 2018h). The STA embodies the central values of its organisation, namely visions and innovative solutions of sustainable transportation and urban 'livability' (ITDP 2018j). By subtly equating BRT with sustainable transport in general, materialised in reduced air pollution emissions and improved mobility for low-income residents, ITDP marginalises other public transport systems that might have similarly sustainable effects. For instance, the former CEO of ITDP lauded Dar es Salaam's sustainable transport measures at the STA ceremony in Washington DC as follows:

> Dar es Salaam's sustainable transport corridor is a watershed moment
> for East African cities. Citizens for the first time are experiencing truly

dignified, safe, healthy experiences as they walk, cycle, and take public transit. Women often outnumber men on the BRT, which serves markets, job centers, schools, neighborhoods. It's the lifeblood of the city.

(ITDP 2018f)

Each year, the previous winning city hosts the *Mobilize* summit, where ITDP announces the next STA winner. Because the summit brings together global transport professionals and researchers and offers them 'the opportunity to experience the city as a learning lab with lessons on how to get world class projects implemented' (ITDP 2018h), it is a prestigious event for the hosting city. Thus, STA winners and global BRT proponents mutually benefit from the strategic combination of STA and *Mobilize*.

Translation

Global forms like facts, standards, models and concepts can travel well due to their strong inherent character of being transferable to different contexts. However, because translation is intrinsic to movement and transformation, global forms are only partially predictable; changing according to their context, they cannot be taken for granted (Morgan 2011; Mukhtarov 2014; Richter 2005). Each arrival brings new mutations, and thus each territorialisation diverges from (global) forms and plans: 'In such processes of translation, new meanings were created and ascribed to activities and experiences. In each new setting, a history of earlier experiences was reformulated in light of the present circumstances and visions' (Sahlin and Wedlin 2008: 224–225). Because experiences flow back from individual systems to the model, I assume that each formation of BRT may potentially impact on the global model. The mutability of the global model expresses itself in the new foci of ITDP's agenda, reflected in the updates of the mobilisation tools *BRT Standard* and *BRT Planning Guide*. Thus, the 'contradictory nature of policies' (McCann and Ward 2011) – being simultaneously relational and dynamic, fixed and territorial – contributes to the continuous de- and reconfiguration of the global form.

Each BRT system is different according to its local context. Designs are altered along the way. Dedicated lanes, as one of the *BRT Basics* and one of BRT's major distinctions from minibus systems, emerge in a variety of materials, colours and dimensions. As DART has shown, translation happens all the time. Not only during global circulatory processes, but also during assembling the global form in Dar es Salaam, the design of DART underwent numerous translational loops. Early in its implementation, DART's physical design of Phase 1 underwent only minor changes, such as a slight reduction of the number of depots and stations due to cost overrun and political disagreements concerning their location (MN 03/2015; PF 10/2015; RL 10/2015). By contrast, the operational design has been mutating unexpectedly strong and extended for global BRT planners (see Chapter 5).[2]

Oscillating between ideals from consultancies, expectations from the Tanzanian government and conditions of the World Bank, the service provider model of DART has existed in different shapes throughout the process of translation.

The idea of replication

Many policies are labelled 'replica' and 'copy' even if scholars and planners have long agreed that literal one-to-one replication is not possible. Mobile policies do not just travel across a landscape; 'they are remaking this landscape' (Peck and Theodore 2010: 170). Policies move in bits and pieces rather than in complete packages, and they '"arrive" not as replicas but as policies already-in-transformation' (ibid.; see also Merry 2006: 48). A contributor to the *BRT Planning Guide* from 2017 states that each BRT system is so unique that it can be seen as a prototype (ITDP 2018c). Another BRT consultant emphasised the local specificities:

> One can't really transfer or copy-cat, carbon-copy-like. Rather, the best a city can hope to do is adapt certain elements to local circumstances; obvious factors like institutional capacities, levels of wealth, local cultures, strength of local unions and incumbent operators, topography, weather, et cetera limit the transferability of experiences from one place to another.
>
> (TG 12/2016)

Nevertheless, central representatives of the BRT community described BRT systems in general as 'replicas' of *Transmilenio*. In Dar es Salaam, DART's physical infrastructure in particular seems to be a direct copy from Bogotá. The former CEO of the DART Agency described how they 'borrowed' the concept from *Transmilenio*: 'All is similar and a replica of Bogotá, including infrastructure' (YK 10/2015). According to her, ITDP encouraged the Brazilian designer to apply 'the Bogotá style' to Dar es Salaam. A DART board member declared: 'What is unique about Dar es Salaam: I don't know. Well, it is more or less copied from the Bogotá model, but with some modifications' (NT 10/2015). Also, a French transport modeller who had contributed to the initial design stated: 'Even if the station is not exactly the same, but the concept for the dimensioning, this is Bogotá' (LL 11/2016). *Transmilenio*'s physical design, particularly the station design, has served as best practice for BRT systems worldwide, including DART and *TransJakarta* (see Matsumoto 2006). The Brazilian engineering office that assembled the initial design for DART in collaboration with a Tanzanian engineering office was inspired by the Colombian best practice. The Tanzanian office translated the initial design into actual conditions by adjusting and adding fundamental parameters like gradients and maximum road width (RH/KD 10/2015). The Brazilian design had three main difficulties.

First, the design did not take into account gradients, as an engineer of the contractor problematised:

> There were problems with the design, some radical problems. For instance, the road was sloped, was going up a hill, and they were putting a bus station on it. You couldn't build a bus station on an angle and if you build a bus station like that, the buses, the doors won't come level with the building because the building, the road slopes this way, the building this way. So, we had quite big adjustments to do in the longitudinal profile.

(WG 10/2015)

Second, all underground elements were absent from the design. DART and its environment had only been depicted on the surface, as two road engineers explained (MN 03/2015; PF 10/2015). The construction companies did not know how deep they could drill before hitting water and drainage pipes. Especially at Kisutu station, where soils were unexpectedly weak, the construction of the concrete lane damaged leaky water pipes just below the surface (see Figure 3.3). Third, the design was not completed within the contract period. Especially components necessary for operations, like the

Figure 3.3 Reopening the concrete surface due to a water leakage (author's photo 05/2016).

station turnstiles or the interfaces of stations and buses for the intelligent transport system (ITS) were only added at later stages.

From plans to practice

Assembling requires translations and trials, and it is not always smooth. Accordingly, DART has undergone numerous translations as it has moved from a global model, to plans and finally to practice. An example of this is the iterative process through which DART's capacity has been calculated. Initially, projected passenger demand was calculated based on globally circulating figures for passengers per hour per direction (pphpd) and then converted into routes and a number of buses. However, shortly before the launch of interim services, international consultants argued that the operational plans did not reflect the actual demand. In order to find out what would need to be adjusted, they assumed that DART 'just had to start' (HM 10/2015). The resistance to the World Bank's international PPP model led not only to changes of the plans and to deviations from the global model, but also to seemingly mundane deviations. Ureta has shown that a script can only represent the transport users' body in the context in which it was developed. Transferred from European bus standards to Santiago de Chile, the script of the handgrips became something different. Because the handgrips were too high for most Chilean passengers, European standards made travelling on *Transantiago*'s buses unsafe and uncomfortable (2015: 103–104). Hence, rather than undermining global forms and standards, allowing scripts to mutate and adapt to local circumstances is necessary.

In the case of Dar es Salaam, a comprehensive process of translation was needed to develop the operational expertise of UDA-RT's drivers, mechanics and coordinators of the control centre. UDA-RT staff travelled to Belgium and Israel to attend workshops on the ITS. Engineers from China instructed them on how to drive, maintain and repair the buses. Since the buses were completely new to Tanzanian mechanics, they could be repaired only at the Jangwani depot, where the mechanics had been trained by the Chinese bus manufacturer. Engineers from Belgium, Germany and South Africa provided information on gear box maintenance and the software of the automated fare collection system (AFCS). Training drivers and mechanics also involved several steps of linguistic translation: because the Chinese mechanics spoke neither English nor Kiswahili and UDA-RT staff spoke no Chinese and only a little English, a Chinese principal of the company translated from Chinese into English, which Tanzanian principals of the Vocational Education and Training Authority (VETA) and UDA-RT then translated into Kiswahili. Training required quick comprehension on the part of the driving instructors and learner drivers, most of whom had never driven an articulated bus. Driving instructors barely had time to learn how to drive the buses before they had to instruct future DART drivers. As one driving instructor summarised, after only one week of theory and one to two weeks of practical training on the BRT corridor, they had to achieve a change of 'Attitude, Mind-Set and Behaviour' (EM 10/2015). This involved learning to drive an 18-m bus, approach the stations properly, give way to non-motorised road users at pedestrian crossings, and brake and accelerate smoothly. Particularly challenging was turning the bus in the narrow turning bay at Kimara Terminal, where the concrete barrier left no margin of error. Here, discipline was essential. The future drivers had to internalise the behavioural ideals of DART services – 'Comfort, Safety and Customer Care' – and modify their previous driving behaviour accordingly. The most challenging task was to restrain themselves on the generally free corridor to the operational speed limit set by the Tanzanian government. Apart from safety concerns, a maximum speed was needed for buses to run according to a timetable – at least in theory.[3] Often, the driving instructors made their students aware of the speedometer so that they could control the speed themselves. Drivers tended to underestimate their speed because they had never driven a new bus with such a powerful but silent engine, and they were not used to having a functioning speedometer.

Inspired by *Transmilenio*, DART can be regarded as a 'bundle' of ideas, life and work cultures, specific techniques and legal arrangements (see Healey and Upton 2010). DART is both very international and highly context specific. On the one hand, the project has emerged out of study trips and visits from global technocrats, as well as dominant documents and globally acknowledged best practices, and the system has been equipped with materials, expertise and regulations from diverse actors and sites. On the other hand, the political economy in Tanzania, the climate conditions in Dar es Salaam, the heritage of daladala practices and other circumstances have contributed to shaping this assemblage. In addition, BRT as a model and its materialisations in specific sites reflect the emergence and institutionalisation of a new form of mobility. BRT's technology and its inscribed ideals and practices have numerous effects and consequences, both on people's behaviour and on urban space. DART differs fundamentally from previous ways of moving in the city. The physical infrastructure promises comfort, reliability and speed; and it stands for ecological and economic sustainability. Materialities legitimate certain political decisions because they create new relationships, practices and discourses. DART's materialities and practices have an effect upon urban life in Dar es Salaam: the overtaking lanes, preventing BRT buses from lining up at stations, and the buses' engines, which comply with low emission standards. Thereby, all components of DART – for example the height of buses and platforms – have to match. Tanzanian laws needed to be combined with the World Bank's regulations regarding compensations, resettlements and salaries. Most importantly, the mismatch between the finalised physical infrastructure and the unclear operational model led to misuse the corridor and stations. In this sense, the line that planners draw between the physical and the operational design does not exist in practice. Many components cannot be classified as either physical or operational, and all components influence each other. DART exists only in the interplay of those different components.

Mwendokasi

Names can take on a range of symbolic meanings, and several meaningful names are prominent in the global BRT sphere. The BRT in Cleveland is called *HealthLine*, in Pune *Rainbow* and in Ahmedabad *Janmarg*, meaning 'People's way' in Hindi. In Johannesburg, the BRT bears the Sesotho words, *Rea Vaya* – 'We are going', similar to Tshwane's BRT *A Re Yeng*, meaning 'Let's go'. *TransBay, TransCaribe, TransCarioca, TransJakarta, Trans Jogja, Transmetro, TransOeste, TransOlimpica, Transantiago* and *Trans Sarbagita* directly refer to their source of inspiration, *Transmilenio*. The *trans-* prefix marks a shared identity, like a family name. However, in Dar es Salaam, *TransDar* did not win through. DART has many names: *Mwendokasi* (high-speed bus), *mabasi yaendayo haraka* (rapid buses), *Dati* (Kiswahili pronunciation of DART), *mabasi makubwa* (big buses) and *mabasi marefu* (long buses). People have different understandings of who and what DART is and whether the meaning of 'rapid' is actually reflected in DART's operations. This deviation is the main reason why the name DART has not caught on. Over time, *mwendokasi* has become the most common label. Transport actors from the DART Agency and UDA-RT criticise that this name is misleading and want to reeducate people to use *mwendoharaka* (fast speed buses): 'We are "mwendoharaka". It is not very fast and not slow since we use our own lane. If you say "mwendokasi" it means these electric high-speed transport means' (GV 09/2016).[4] The CEO explains that DART is *haraka*, but not *kasi*, with its average speed of twenty to thirty km/h. *Mmwendokasi* is inappropriate because it means 'speed' and not 'rapid', and hence misleads citizens about the new bus service (GV 09/2016).

The meaning of *mwendokasi* is nonetheless overall positive. It has gained a growing political dimension under the politics of President Magufuli, and also as a term in popular culture (e.g. the *Mwendokasi Festival* in Dar es Salaam). *Mwendokasi* has the sense of doing things efficiently and is associated with attempting innovation or modernisation. It indexes a Dar es Salaam that is speeding up, leaving behind the famous Kiswahili motto: 'Hurry is not a blessing, slowness is the way to go'.[5] The Tanzanian journalist Mngodo (2016) describes her experience of a DART trip by comparing the velocity of DART with the general speed of urban life in the city:

> Facebook does make it hard, when everyone is showing off their life that's moving at the speed of a 'Dart bus' and there you are as slow as a daladala. [...] It took us exactly 40 minutes to reach the end of the journey; a travel that would usually take 2–3 hours depending on traffic.

This was amazing and strange. I wasn't used to being on time, at least not while taking a bus.

Hence, DART has become the symbol of rapidness. Urban life in Dar es Salaam benefits from the BRT system, as it accelerates the city through its name and image, its materialisation, agency and practice. DART's environment needed to assimilate to the new speed. Not only passengers like Mngodo but also DART's drivers needed to develop an understanding of what exactly this rapidness means. They had to learn that the 'rapid' in BRT and DART does not refer to the maximum speed but to the average operational speed.

The rapidness of the DART project needs to be qualified in two respects. First, since DART's assembling was deferred, the system materialised slowly, much slower than expected. BRT consultants claim that a process from initial planning to operations of one BRT phase can be done in less than three years (JM 03/2015; PH 03/2015; Wright 2003: 4). However, estimates for DART differ drastically – from six to twenty years – regarding the time needed to implement the remaining five project phases. A global BRT consultant described DART as 'the slowest BRT in the world' (LL 11/2016). Moreover, the deferrals of the construction period slowed down the regular traffic longer than initially expected on Morogoro Road, the major axis of the city. This affected the city's traffic as a whole because vehicles had to use other roads. Even if DART has led to less traffic on mixed-traffic roads, the system's priority in traffic simultaneously constrains the mixed-traffic. For example, right-turning vehicles at Morogoro Road and Kawawa Road, which have to cross the BRT corridor, have much fewer options, lengthening journeys of regular traffic both in time and distance. Second, DART operations themselves do not run as rapidly and smoothly as intended. During heavy rains, a normal feature of Dar es Salaam's climate, BRT traffic regularly comes to a halt due to inundated roads (see Chapter 6). In addition, insufficient scheduling due to the incomplete installation of the control centre and insufficient bus numbers leads to spatially unequal deployment of DART buses, as messages and photos by station attendants regularly illustrate in the WhatsApp group of DART's stakeholders (see Chapter 2). The congestion of DART buses at some stations, while other stations are overcrowded with passengers waiting for a bus, shows that uncoordinated services continuously slow down both buses and passengers.

Standardisation

The *BRT Standard* is the latest central tool created by ITDP. It has had an impressive career since its first publication in 2012, making its way to various locations in the Global South and influencing ideas of how a high-quality BRT system should look. According to ITDP, the *BRT Standard* creates consensus about BRT and 'defines and refines best practices, and provides metrics for evaluation' (ITDP 2020). This tool is thought to make BRT less mutable, to stabilise it. The concept of BRT is supposed to establish itself globally by excluding so-called simple bus systems and alternative descriptions for so-called advanced bus systems (e.g. busway, bus priority system). The *BRT Standard* is the tool that requires the most strictness and the least flexibility: tables show the correlation between BRT systems with PPP arrangements and the *Gold Standard* of BRT systems, surveys show a decrease in travel time of BRTs with centre-aligned stations and overtaking lanes, and stories tell of increased road safety when pedestrian crossings are installed instead of pedestrian bridges. Through this self-referential strategy – assessing BRTs according to the evaluation system that the NGO created – ITDP institutionalises the *BRT Standard*, its inscribed ideals – and itself.

Standards have become a central feature of political, social and cultural life (Lampland and Star 2009: 10). They define, restrict, compare, order, grade, (de)valorise, exclude and rank. According to Bowker and Star, standards classify the world:

> A 'standard' is any set of agreed-upon rules for the production of (textual or material) objects. [...] A standard spans more than one community of practice (or site of activity). It has temporal reach as well in that it persists over time. [...] Standards are deployed in making things work together over distance and heterogeneous metrics.
>
> (2000: 13–14)

Standards are inextricably linked to travelling models because they aim to facilitate the 'reproduction of actions in different places' (Guggenheim 2016: 67). Travelling in time and space, standards aim to construct robust, objective realities; and they enable coordination and cooperation over distance. Standards are a tool to perform the world through technical practices (Busch 2011: 68–74, Mol 2002). Standards themselves become powerful because they have *'the ability to set the rules that others must follow, or to set the range of categories from which they may choose'* (Busch 2011: 28, emphasis in original). The normative power of standards provides them with agency to produce hierarchical relationships, and to shape global processes (Brunsson and Jacobsson 2000; Strathern 2005: 466). Although standardisation implies legibility and hierarchy, its primary goal is not to necessarily improve quality but to bring consistency, similarity and homogeneity. Standards aim to harmonise the design of transport systems, the size and taste of vegetables, security settings for software applications, et cetera. Powerful institutions define a norm to which actors need to comply if they want to avoid being marginalised and even excluded.[6]

The *BRT Standard* embodies and materialises hierarchic relations and normativities of the global BRT model. The tool excludes some bus systems from the global BRT model while giving others an advantage, materialised in high scores and the labels of *Bronze*, *Silver* or *Gold*. Standards can be of different types and have different meanings. For instance, using 'gold' and 'silver' as descriptors directly evokes the idea of a guaranteed value (Busch 2011: 22). The *BRT Standard* is based on a point system. First, it assigns points to BRT systems, and second it 'scores' – in ITDP's words – labels from *Basic* to *Bronze*, *Silver* and *Gold*, according to the amount of points. The two outstanding certified *Gold Standards* – *TransOeste* in Rio de Janeiro and *Transmilenio* – 'inspired a wave of BRT innovation around the world' (ITDP 2014b). BRT systems are not scored as a whole, but per corridor, since the qualities and characteristics of several corridors of a system might vary (ITDP 2016a; PH 12/2016). In 2016, ITDP had fully scored more than one hundred BRT corridors in approximately sixty cities, which corresponds to roughly two thousand kilometres (ER 08/2016; ITDP 2016b).

The *Basics* of the *BRT Standard* define BRT at a minimum level, distinguishing BRT from so-called regular buses. ITDP claims that for a bus system to be officially called BRT, it needs to have the following characteristics: dedicated right-of-way, busway alignment, intersection treatment, off-board fare collection and platform-level boarding. ITDP declares that those 'essential five elements, *The Basics*, are basically putting the "rapid" in "rapid transit"' (ITDP 2018j). A BRT corridor with less than 28 (out of 100) points is not considered BRT. In order to achieve a score above *Basic*, BRT corridors need to fulfil a number of criteria regarding the design of the physical infrastructure, integration and access, service planning, branding, and passenger information. The NGO implemented the *BRT Standard* because it strived for a common definition and a tool for quality control in order to prevent the 'backlash against BRT' by 'modest bus systems' (ITDP 2014b). ITDP member Ethan Robinson, responsible for the latest edition of the *BRT Standard*, explained:

> There was such disparity about what people actually considered to be BRT, so much that we got to the point that we were saying: 'BRT really didn't mean anything.' [...] The term was very confusing to a lot of people and people didn't trust it as kind of a brand. And so, unlike other things like metro or even light rail, BRT suffered from having an image problem. [...] Then, we decided that we need to have a more of a global standard of what BRT was.
>
> (ER 08/2016)

Since then, a technical committee has updated the tool three times (ITDP 2018a). Like the *BRT Planning Guide*, it is published online as an open access document. Unlike the *BRT Planning Guide*, it is still available in Bahasa-Indonesia, Chinese, English, Portuguese and Spanish. The central aim of the *BRT Standard* has not changed significantly over the years. The last update in 2016, conducted in conjunction with the update of the *BRT Planning Guide*, differs from earlier versions primarily in regard to its increased number of pages and content, slight changes in the methodology, and some adjustments to the categories Safety and Operations. Robinson described its assembling as follows:

> It's an on-going process and it's never going to be perfect because it's not a science. This is achieving agreement among people who think about this and care about BRT to – if not necessarily agree on all the details but at least – to come to a point where they're willing to put their name on a document that says: 'This is what we, a group of experts, think is good and helpful for understanding what BRT is.'
>
> (ER 08/2016)

ITDP's continuous creating, strengthening and protecting BRT as a brand thus mean that regular bus systems cannot carry the BRT label if they do not comply with the criteria given. On the assumption of global applicability, the *BRT Standard* supports the rapid mobilisation of BRT as the only sustainable transport solution for the Global South. Another ITDP member talked to me about 'moving in the right direction', referring to the current 'inflection point' of African cities orienting their BRT plans to ITDP's criteria in the *BRT Standard* (PH 11/2017).

Importantly, the *BRT Standard* itself is a product of global policy circulation. ITDP was inspired to compose the standard from the Leadership in Energy and Environmental Design, a tool from the US Green Building Council that claims to be 'a globally recognized symbol of sustainability achievement' (2018). The tool rates green buildings based on certification points as *certified*, *silver*, *gold* or *platinum*. Both tools calculate a weighted sum that generates a comparable hierarchical structure. ITDP's tool provides guidelines rather than rules. Because it is not legally binding, the *BRT Standard* is more flexible than widespread standards like EU norms on meat production (see Dunn 2005). However, the NGO tries to naturalise its tool in order to increase its global effectiveness:

> [...] Standards, like the world of nature, are seemingly 'supposed' to be the way they are. What could be more powerful than something that is revealed as no less than a part of the natural world itself?
>
> (Busch 2011: 32–33)

Using the *BRT Standard* as source for inspiration in the planning phase is part of ITDP's agenda, and the NGO uses every occasion to invite city governments to 'MAKE IT A GOLD!!!' (ITDP 2014a). Moreover, ITDP asserts that it rewards BRT corridors by scoring them, instead of denigrating unsatisfactory BRT corridors. The tool is supposed to create incentives for cities that consider the implementation of best practice BRT (ER 08/2016). Robinson emphasises that these incentives should be in line with the respective possibilities. For instance, cities with limited budget should better go for a proper *Bronze* than for a half-hearted *Gold*. ITDP is interested in producing as many high scores as possible, since this reflects its own work as a successful consultant. In nearly every category, ITDP applies ranges that offer each BRT corridor the possibility to earn points even if it does not apply best-practice designs. For instance, the points for off-board fare collection are weighted percentagewise: four to eight points are awarded according to the percentage of the stations on the corridor with barrier-controlled fare collection, routes using the corridor bus infrastructure for proof-of-payment, and routes using the corridor bus infrastructure for on-board fare validation at all doors (ITDP 2016a: 35). ITDP emphasises that certain characteristics depend on the context. Concerning station dimensions, the NGO sets minimum length and width, but it adds that local circumstances like climate conditions ultimately define how well a station is designed. In the *BRT Planning Guide*, ITDP illustrates this approach with a quote of the musician Kanye West: 'I refuse to accept other people's ideas of happiness for me. As if there's a "one-size fits all" standard for happiness' (West in ITDP 2017b: 837). However, since context-specific assemblings are not taken into account in the calculation of the corridors' scores, West's assertion becomes irrelevant.

Full versus lite

BRT consultants distinguish between 'full BRT' and 'BRT Lite'. The main difference is the location of the BRT lane and the stations, which are either median-aligned or kerbside-aligned (Ferbrache 2019: 3). A major concern of ITDP is to realise so-called full BRT systems instead of BRT Lite systems as they exist in Yangon or Lagos (LAMATA 2017; Mobereola 2009; Orekoya 2010). According to the organisation, BRT Lite does not satisfy the vision of sustainable and high-quality BRT. Whereas full BRT systems generally have lanes and stations located at the centre of the road, lite systems tend to locate them on the outer lane. According to ITDP, this kerbside-location is reminiscent of 'an old bus on a bus lane' (ITDP 2018d).

> ITDP gives a workshop on NMT-friendly street design on the second conference day of the conference of sustainable urban transportation in Nairobi. Members of the NGO guide the separate workshop groups and politely persuade the participants to focus on the inclusion of median-aligned BRT in the street design. Median alignment does have advantages for non-motorised transport – however, the focus shifts on (full) BRT, whereby the task officially did not even provide the need to include BRT in the designs at all. In the subsequent evaluation, it further turns out that ITDP has indirectly linked the workshop to actual BRT plans for one of Nairobi's major roads, whose development ITDP had been advising on for years. The day before at the conference, another international transport consultancy called the BRT plans for this road into question. (November 2016)

In order to clearly distinguish BRT from a simple bus system, the creator of the *BRT Standard* gives preference to median-aligned BRT corridors, which are marked with a green tick on slides shown at conference presentations and webinars. According to the *BRT Standard*, a median-aligned BRT lane earns up to eight points, whereas alignment at the kerbside only earns a maximum of three points. Consequently, BRT Lite can achieve the status of a *BRT Basic* but can only barely achieve the next level, *Bronze*. ITDP justifies its disapproval of BRT Lite with operational arguments: median alignment ensures BRT services with few obstructions, minimises operational delays and prevents conflicts with mixed traffic. Service vehicles need the outer lane for short stops, and right/left turns on mixed-traffic lanes need to cross the outer lanes – i.e. the BRT lane located at the kerbside. Hence, ITDP has created a tool which distinguishes 'true BRT' not only from 'regular bus systems' but also from 'BRT Lite'.

As both the creator of the *BRT Standard* and the global proponent of high-quality BRT, ITDP has a double role. On the one hand, ITDP consults cities on BRT designs and funding sources. On the other hand, it decides on the *BRT Standard*'s categories and weighting, and chooses if the assessment of a corridor is published. ITDP is the biggest advocate of the *BRT Standard*, promoting it to the global crowd, presenting it as a universally, i.e. global and all-encompassing, applicable tool. However, only its creator officially acknowledges this self-referential tool and increasingly showcases its label on Africa:

> ITDP played a pivotal role in the implementation of Johannesburg's silver-standard Rea Vaya BRT corridors and in Cape Town's bronze-standard MyCiTi BRT, and continues to support the expansion of both systems. ITDP was also instrumental in starting and supporting the development of Dar es Salaam's BRT, now under construction, and set to open in 2015 as Africa's first gold-standard BRT.
>
> (ITDP 2018k)

By referring to the standards for corridors planned (DART) or achieved (*Rea Vaya* and *MyCiTi*), ITDP references itself, thereby (allegedly) proving its great influence in urban transport planning and its successful project implementation.

However, the dominant tools and inscribed global values do not necessarily have positive outcomes, as experiences from the two South African cities show. In Cape Town and Johannesburg, ITDP's so-called benchmarking system led to the cities devoting more attention to the *BRT Standard*'s criteria than to their own transport needs (Wood 2015: 1075). The case of DART indicates rather the opposite. When the Dar es Salaam City Council started plans for DART in collaboration with ITDP and the World Bank, the *BRT Standard* did not even exist. However, something surprising occurred, like a time shift or an anachronism.

On the ITDP website, Dar es Salaam's plan of the early 2000s is described as 'Africa's first gold-standard BRT' (ITDP 2018k). However, the description 'Africa's first gold-standard BRT' stands in sharp contrast to actual circumstances in Dar es Salaam (see Chapter 6), where the *BRT Standard* has yet to play a significant role. The majority of DART's key actors were not even aware of its existence, and the few internationally embedded actors who did know about it did not take the ranking seriously: 'The ranking of *Silver*, *Gold* or whatever is not important. But what is important to us, are those indicators of quality' (JK 10/2015).

It is likely that the tool will continuously increase its global impact due to the generally strong position of ITDP as promoter of this tool, and the absence of an alternative evaluation tool for standardising this transport model. According to several ITDP members, a long list of cities are waiting to be scored. The broad public interest in the *BRT Standard* materialises in the high number of downloads of the pdf document. The actual impact of this tool is hard to measure though. ITDP is uncertain whether cities hope to benefit from their score, and if they use it for city-branding:

> We have seen that but not as much as we would want. But yes, we definitely have seen cities brand themselves as having *Gold Standard* BRT. I don't want to overstate how often that's done. [...] Albuquerque has really pushed that they're building a *Gold Standard* BRT. I don't know if there're other good examples where they have actually promoted it. The answer is: not really. But, you know, there have been some. And I think that Chicago is also promoting that they want to build a *Gold Standard* BRT.
>
> (ER 08/2016)

The few BRT technocrats who apply the *BRT Standard* are linked to ITDP in some way. Trevor Gabriel, a member of the *BRT Standard*'s technical committee, referred to the tool in order to see 'who has been best at what' (TG 12/2016). Institutions that financially support the *BRT Standard* endorse it by using it as a checklist to determine which elements are indispensable for BRT. Actors that financially support the realisation of BRT systems – like the World Bank that grants loans to governments in the Global South – use the *BRT Standard* to 'push those things' because they have been 'identified by experts' (ER 08/2016). The notion of pushing again demonstrates the self-interest of ITDP. The NGO sees its potential growing in accordance with a rising global popularity of the *BRT Standard* and a rising number of high-quality BRT systems. The *BRT Standard* is performatively effective; instead of being an objective tool that quantifies the operational performance of BRT corridors, it decides which corridors count as BRT and which label they carry. The *BRT Standard* may well influence decision making in the planning phase and the materialisation of plans. The tool, moreover, increases the similarities of urban transport in cities of the

Global South. Despite its lack of legislative embeddedness, it has been become rapidly entrenched and pervasive since 2012.

Rigidity and flexibility

ITDP aims to determine the success of its own projects through its presumable rigid and objective *BRT Standard* – an endeavour that largely succeeds. Officially, external transport consultants (who are mainly from the US, Europe and Latin America), are gathered in technical committees and go to the sites to score them. In order to prevent bias, committee members are not directly linked to ITDP, at least not officially. The need for external consultants raises questions about the evaluation method and categories of the tool: shouldn't a standard be non-negotiable, contextually unbound and transferable to anyone and anywhere? In my experience, it is not necessary to be a transport planner to conduct the scoring; some knowledge of the corridor to be scored and a copy of the publicly available, self-explanatory spreadsheet used to calculate the points is sufficient. However, categories and points are ultimately negotiable. ITDP has its own ways of contributing to, and interfering in, the evaluation. Not only does it decide what is published, it also directly intervenes in the evaluation, as my experience regarding the scoring of DART reveals:

> Independently from one another, the CEO of the DART Agency and I score DART's Phase 1 half a year after the start of operations. The CEO does the scoring at his own initiative, and I score DART during my internship with ITDP as a preliminary version for the official scoring to follow by the technical committee. Both of us come out with a low *Bronze* score. More than twenty points are missing for a *Gold Standard*. After revising my preliminary scoring with ITDP's Africa programme director and his more generous gaze, we upgrade DART's score to a low *Silver* – better, but still several points away from the intended *Gold*. DART still receives point deductions in the categories Communications, Access and Integration, and Operations.
>
> (December 2016)

In addition to the twisted temporality of DART's golden forecast, this prediction would be hard to realise based on its actual performance. Given the way that ITDP has been globally promoting DART, I expected the NGO to prioritise the scoring of DART's Phase 1. However, as of mid-2020, an official scoring of DART is yet to be completed. Is ITDP avoiding publishing DART's (actually lower than *Gold*) score? Or is it just giving DART a lot of time to improve before doing the official scoring, so that it still has a chance of reaching *Gold*? ITDP generally suggests scoring corridors six months after launch, assuming that BRT operations need this long to settle.

The ridership has time to grow, and operators have time to become used to managing operations with the ITS:

> They should be pretty good by then. But if they are not, that's a decade of bigger problems. [...] The point of the *Standard* is to score things as they are. And so, if they have a corridor that doesn't have a control center after six months when you do the scoring, then you score without the control center.
>
> (ER 08/2016)

This appraisal would mean that DART was projected to have enduring problems without the control centre fully installed after six months. Nevertheless, ITDP and its peers continue to present DART as *Gold* (Hook in ITDP 2018e), which means that ITDP uses its own tool in order to improve DART's image as well as its own consultancy work. The case of DART further illustrates the central problem of applying a classification tool globally. The *BRT Standard* does not consider special cases like the DART interim service, which has been operating with a reduced bus fleet and a minimal use of the control centre for several years (see Chapter 5). The *BRT Standard* allows for a rescoring if the corridor has undergone significant changes so that systems have the chance to improve their image. This shows that the *BRT Standard* can be more flexible and generous than it seems at first sight.

The *BRT Standard*'s objectivity and rigidity can be further called into question in two ways. First, the notions of 'world class' and 'best practice' appear frequently in ITDP's publications, referring in most instances to BRT systems that have received *Silver* or *Gold* (ER 08/2016; ITDP 2017b: 389). This indicates that ITDP endeavours to equate these prestigious terms with the *BRT Standard*. In turn, this means that DART can only be presented as a best practice or world class BRT if it is considered of *Silver* or *Gold Standard*. Second, ITDP admits that it disproportionately considers the better corridors because the NGO does most full scorings of BRT corridors that are more prominent or that are expected to receive comparably high scores. Remarkably, all corridors evaluated have reached the *Basics* score, which means that all corridors scored can be considered as BRT. This result stands in contrast to ITDP's overall goal to distinguish 'true BRT' from 'regular bus systems'. If all BRT corridors were actually scored, not all of them would receive the label of (at least) *BRT Basics*. ITDP tries to balance between the claim that the *BRT Standard* is 'not about pushing anyone to a particular vision of what is BRT or what is not BRT' (ER 08/2016), and the aim of enforcing uniformity and specific quality standards like 'world-class passenger experiences, significant economic benefits, and positive environmental impacts' (ITDP 2012, 2016a).

Standards beyond the Standard

Beyond the *BRT Standard* itself, standardised BRT planning is articulated in a variety of presumably universal numeric forms, benchmarks, formulas and ratios, such as optimal kerbstone height, maximum speed, distance between stations, highest pphpd and lowest construction cost per kilometre. Numbers and standards help to uniformly measure and label, include and exclude transport systems, and are a prerequisite for distributing and improving the global BRT model. Numbers and standards have something in common with Latour's notion of inscriptions. Not only can they be made mobile and immutable, they are also optically consistent:

> Because of this optical consistency, everything, no matter where it comes from, can be converted into diagrams and numbers, and combination of numbers and tables can be used which are still easier to handle than words or silhouettes.
>
> (1986: 20)

As an essential part of (transport) planning, standards are created in order to produce functionally comparable results in different contexts, e.g. making different BRT systems comparable (c.f. Collier and Ong 2005: 11). In heterogeneous domains, objects and global forms gain legibility and functionality through those standards (Collier 2006: 400). Thereby, standards produce knowledge and stabilise (new) ideas as facts:

> Standards function [...] as 'fact factories'. Not only do they import knowledge about how things should be made, but also, by specifying particular forms of data collection, recording, and analysis, they act as engines for generating knowledge about products, processes, and people.
>
> (Dunn 2005: 184)

On an ITDP webinar, one of the global BRT protagonists admitted that the practice of constructing a BRT system is much more complex than can be captured by the formulas. He stated that transport planners and politicians need to take additional aspects into account, and he declared that the formula for calculating traffic demand and fleet size could only serve as a rule of thumb (ITDP 2018b). However, BRT proponents tend to convert stories into representative numbers because numbers circulate more smoothly than words.

Mutation takes place in translational processes. Robinson from ITDP questions the global applicability of this tool, at least to a limited extent:

> A lot of the debate was centred around whether you can have one document that creates a standard for everywhere in the world – because of

the different realities on the ground in different places. [...] How do you capture all those things in one single standard? Honestly, the *Standard* is never gonna be perfect, but the idea was to make some sort of comparison across places, different places around the world, using the same methodology.

(ER 08/2016)

ITDP's goal of controlling the travelling model through the *BRT Standard* shows similarities with EU norms on food processing and with methods of European long-distance control (see Dunn 2005; Law 1986). Standardising BRT is a central, self-referential tool that ITDP uses to control BRT and bring the model to (new) market(s). Investigating the *BRT Standard* reveals that ITDP attempts to create a BRT model that is immutable and mobile. According to Busch's (2011) differentiated understanding of standards, the *BRT Standard* definitely counts as a standard. However, it is more mutable than most legally acknowledged standards, and can rather be regarded as a label (e.g. a fair-trade label) than an official standard (e.g. ISO or EU norms). The changing shape of the *BRT Standard* materialises in space and time, in regular revisions and context-specific translations.

By looking closely at BRT's mobilisation, translation and standardisation, this chapter has shown that the global BRT model is fluid and mutable, but that it also has rigid and immutable characteristics. Understanding translation as the product of displacement and transformation, things permanently change and adapt to new circumstances. The BRT model resembles the bush pump with its fluid boundaries (see De Laet and Mol 2000). Turning Latour's 'immutable mobiles' (1986, 1987) around, the model is only mobile and successful because its inscribed metrics, knowledge and values are modifiable – to a certain extent. In order to be disseminated globally, the model needs to be simultaneously mutable and immutable, fluid and standardised. It mutates over time – expressed in the regular updates of the mobilisation tools – and across space, e.g. in the varying degrees of generosity of scoring. Facilitated by a broad range of consultants and their tools, the model's adaptability enables its translation from one site to another. Unexpected formations of BRT emerge, deviating from the model, standards and plans. These fluid formations reflect controversies and negotiations on power on the one hand, and ordinary context-specific conditions on the other hand. BRT is a mutable mobile, flowing 'in different configurations into different locations' (Larner and Le Heron 2002: 760). The hybrid model does have a few essential characteristics, which ITDP labels as the *Basics* of BRT. By creating a tool that claims objectivity, ITDP uses the *BRT Standard* strategically to enforce certain urban transport ideals. Institutionalising knowledge and ideals in the *BRT Standard* gives authority to the NGO. ITDP's famous and effective tools – the *BRT Planning Guide* and the *BRT Standard*, the *STA* and the *Mobilize* summit – share the same values. They aim both to increase and control the quality of BRT, and they balance rigidity and flexibility, two essential ingredients for the dissemination of the global BRT model. Therefore, the tools have contributed to a remarkable mobilisation and stabilisation of the BRT model.

Notes

1 BRT mobilisers promise that through the implementation of a 'world class BRT', the status of becoming a 'world class city' may well be reached (c.f. Holzwarth 2012: 35; Paget-Seekings 2015: 119). Narratives of how 'Dar can become world-class BRT' (RL 10/2015) have already been constructed, and ITDP emphasises DART's 'world-class stations' (ITDP Africa 2020a). ITDP moreover describes Nairobi's BRT attempts as a 'world-class BRT project', being guided by 'world-class BRT engineers' (ITDP Africa 2020b).
2 A side effect of the delays of construction and implementation was that the data on which the initial designs were based had become outdated due to rapid urban growth: between 2005 and 2013, traffic demand for Phase 1 had risen by 50 per cent (HM 10/2015; LL 11/2016). Data could not just be updated and adjusted, but had to be collected from scratch.
3 Years later, UDA-RT still operates without timetables.
4 *Sisi ni 'mwendoharaka'. Sio haraka sana na sio polepole kwa sababu tunatumia barabara zetu mwenyewe. Ukisema 'mwendokasi' ni kama vile high-speed ambazo zinatumia umeme.*
5 *Haraka haraka haina baraka, pole pole ndio mwendo.*
6 For more elaboration, see Barry (2015) and Dunn (2005) on the creation of common standards in the EU through the assembling of hierarchical transnational networks. Dunn describes how the post-Soviet Polish meatpacking industry has been forced to apply EU standards on management and food quality (e.g. purity, quality and the value of the pork) in order to be allowed to enter the EU market: 'The object of introducing standards in postsocialist enterprises was not just to increase product quality, but to reshape firms in order to make them more closely resemble the organizational forms of their Western counterparts' (ibid.: 176-177).

References

Barry, A. 2015. "Infrastructural Times". In: H. Appel, N. Anand and A. Gupta (eds.), *The Infrastructure Toolbox*. Accessed at *Cultural Anthropology* (https://culanth. org/fieldsights/series/the-infrastructure-toolbox). Published 24.09.2015, retrieved 17.04.2020.

Barry, A. 2013. "The Translation Zone. Between Actor-Network Theory and International Relations". *Millennium: Journal of International Studies* 41:3, 413–429.

Behrends, A., S.J. Park and R. Rottenburg 2014. "Travelling Models. Introducing an Analytical Concept to Globalisation Studies". In: ibid. (eds.), *Travelling Models in African Conflict Management. Translating Technologies of Social Ordering*. Leiden: Brill, 1–40.

Bowker, G. and S. Star 2000 (1999). *Sorting Things Out. Classification and Its Consequences*. Cambridge, MA: MIT Press.

Brunsson, N. and B. Jacobsson 2000. *A World of Standards*. Oxford: Oxford University Press.

Busch, L. 2011. *Standards. Recipes for Reality*. Cambridge, MA: MIT Press.

Callon, M. 1986. "Some Elements of a Sociology of Translation. Domestication of the Scallops and the Fishermen of St Brieuc Bay". In: Law, J. (ed.), *Power, Action and Belief. A New Sociology of Knowledge?* London: Routledge, 196–223.

Collier, S. and A. Ong 2005. "Global Assemblages, Anthropological Problems". In: A. Ong and S. Collier (eds.), *Global Assemblages. Technology, Politics, and Ethics as Anthropological Problems.* Malden, MA: Blackwell, 3–21.

Collier, S. 2006. "Global Assemblages". *Theory, Culture & Society* 23, 399–401.

Cook, I. and K. Ward 2012. "Conferences, Informational Infrastructures and Mobile Policies. The Process of Getting Sweden 'BID Ready'". *European Urban and regional Studies* 19:2, 137–152.

De Laet, M. 2000. "Patents, Travel, Space. Ethnographic Encounters with Objects in Transit". *Environment and Planning D* 18, 149–168.

De Laet, M. and A. Mol 2000. "The Zimbabwe Bush Pump. Mechanics of a Fluid Technology". *Social Studies of Science* 30:2, 225–263.

Dunn, E. 2005. "Standards and Person-Making in East-Central Europe". In: A. Ong and S. Collier (eds.), *Global Assemblages. Technology, Politics, and Ethics as Anthropological Problems.* Malden, MA: Blackwell, 173–193.

Eisenstein, E. 1979. *The Printing Press as an Agent of Change. Communications and Cultural Transformations in Early-Modern Europe.* Cambridge: Cambridge University Press.

Eriksen, T. 1991. "Walls: Vanishing Boundaries of Social Anthropology". Accessed at *Engaging With the World. Eriksen's Site* (http://folk.uio.no/geirthe/Walls.html). Published 14.09.1991, retrieved 17.04.2020.

Ferbrache, F. 2019. "The Value of Bus Rapid Transit in Urban Spaces". In: ibid. (ed.), *Developing Bus Rapid Transit. The Value of BRT in Urban Spaces.* Cheltenham: Edward Elgar Publishing.

Filipe, L. and R. Maćario 2013. "A First Glimpse on Policy Packaging for Implementation of BRT Projects". *Research in Transportation Economics* 39, 150–157.

Freeman, R. 2012. "Reverb. Policy Making in Wave Form". *Environment and Planning A* 44, 13–20.

Guggenheim, M. 2016. "Im/mutable Im/mobiles. From the Socio-Materiality of Cities towards a Differential Cosmopolitics". In: A. Blok and I. Farías (eds.), *Urban Cosmopolitics. Agencement, Assemblies, Atmospheres.* London, New York: Routledge, 63–81.

Healey, P. and R. Upton 2010. *Crossing Borders. International Exchange and Planning Practices.* London, New York: Routledge.

Holzwarth, S. 2012. "Bus Rapid Transit Systems for African Cities". *Trialog* 110, 32–37.

ITDP 2020. "History of ITDP". Accessed at *ITDP* (https://www.itdp.org/who-we-are/history-of-itdp/), retrieved 17.04.2020.

ITDP Africa 2020a. "Dar es Salaam". Accessed at *ITDP Africa* (https://africa.itdp.org/cities/dar-es-salaam/), retrieved 17.04.2020.

ITDP Africa 2020b. "Nairobi". Accessed at *ITDP Africa* (https://africa.itdp.org/cities/nairobi/), retrieved 17.04.2020.

ITDP 2018a. "BRT Planning 101". Accessed at *ITDP* (https://www.itdp.org/publication/webinar-brt-planning-101/). Published 02.02.2018, retrieved 17.04.2020.

ITDP 2018b. BRT "Planning 201: Service Planning". Accessed at *ITDP* (https://www.itdp.org/publication/webinar-brt-planning-201/). Published 02.03.2018, retrieved 17.04.2020.

ITDP 2018c. "BRT Planning 401: Infrastructure and Design". Accessed at *ITDP* (https://www.itdp.org/2018/05/04/webinar-brt-planning-401/). Published 04.05.2018, retrieved 17.04.2020.

ITDP 2018d. "BRT Planning 501: Integration". Accessed at *ITDP* https://www.itdp.org/2018/10/09/webinar-brt-501-integration/). Published 09.10.2018, retrieved 17.04.2020.

ITDP 2018e. "BRT Planning 601: Business Plans". Accessed at *ITDP* (https://www.itdp.org/2018/11/05/webinar-brt-601-business-plans/). Published 05.11.2018, retrieved 17.04.2020.

ITDP 2018f. "Dar es Salaam, Tanzania Presented with 2018 Sustainable Transport Award". Accessed at *ITDP* (https://www.itdp.org/2018-sta-ceremony/). Published 30.01.2018, retrieved 17.04.2020.

ITDP 2018g. "DART: Transforming Urban Mobility in East Africa". Accessed at *ITDP* (https://www.itdp.org/2018/06/01/webinar-dart-transforming-mobility/). Published 01.06.2018, retrieved 17.04.2020.

ITDP 2018h. "MOBILIZE". Accessed at *MobilizeSummit* (https://mobilizesummit.org), retrieved 11.11.2018.

ITDP 2018i. "The Sustainable Transport Award". Accessed at *Staward* (https://staward.org), retrieved 17.04.2020.

ITDP 2018j. "What Is BRT"? Accessed at *vimeo* (https://vimeo.com/276223930). Published 21.06.2018, retrieved 17.04.2020.

ITDP 2018k. "Where We Work: Africa". Accessed at *ITDP* (https://www.itdp.org/-where-we-work/Africa), retrieved 11.11.2018.

ITDP 2017a. "Dar es Salaam, Tanzania Wins 2018 Sustainable Transport Award". Accessed at *ITDP* (https://www.itdp.org/dar-es-salaam-wins-2018-sta/). Published 02.07.2017, retrieved 17.04.2020.

ITDP 2017b. "The Online BRT Planning Guide". 4th Edition. Accessed at *ITDP* (https://brtguide.itdp.org). Published 16.11.2017, retrieved 17.04.2020.

ITDP 2016a. "BRT Standard. 2016 Edition". Accessed at *ITDP* (https://www.itdp.org/2016/06/21/the-brt-standard/). Published 21.06.2016, retrieved 17.04.2020.

ITDP 2016b. "BRT Standard: Scores". Accessed at *ITDP* (https://www.itdp.org/brt-standard-scores/), retrieved 17.04.2020.

ITDP 2014a. "Bus Rapid Transit for Indian Cities". (*unpublished*).

ITDP 2014b. "The BRT Standard. 2014 Edition". Accessed at *ITDP* (https://itdpdotorg.wpengine.com/wp-content/uploads/2014/07/BRT-Standard-20141.pdf), retrieved 17.04.2020.

ITDP 2012. "BRT Standard. Version 1.0". Accessed at *Infrastructure USA* (https://www.infrastructureusa.org/wp-content/uploads/2012/02/BRT_Standard_12312.pdf), retrieved 17.04.2020.

LAMATA 2017. "BRT Means Bus Rapid Transit". Accessed at *Lagos Metropolitan Transport Authority* (http://lamata-ng.com/brt-bfs/), retrieved 17.04.2020.

Lampland, M. and S.L. Star 2009. "Reckoning with Standards". In: ibid. (eds.): *Standards and Their Stories*. Ithaca, London: Cornell University Press, 13–24.

Larner, W. and R. Le Heron 2002. "The Spaces and Subjects of a Globalising Economy. A Situated Exploration of a Method". *Environment and Planning D* 20, 753–774.

Latour, B. 1999. *Pandora's Hope. Essays on the Reality of Science Studies*. Cambridge, MA: Harvard University Press.

Latour, B. 1994: "On Technical Mediation". *Common Knowledge* 3:2, 29–64.

Latour, B. 1987. *Science in Action. How to Follow Scientists and Engineers through Society.* Cambridge, MA: Harvard University Press.

Latour, B. 1986: "Visualisation and Cognition. Drawing Things Together". *Knowledge and Society – Studies in the Sociology of Culture Past and Present* 6, 1–40.

Latour, B. and S. Woolgar 1986. *Laboratory Life. The Construction of Scientific Facts.* Princeton, NJ: Princeton University Press.

Law, J. 2011. "Heterogeneous Engineering and Tinkering". Accessed at *Heterogeneities* (http://www.heterogeneities.net/publications/Law2011HeterogeneousEngineeringAndTinkering.pdf). Published 14.11.2011, retrieved 17.04.2020.

Law, J. 2002: "Objects and Spaces". *Theory Culture & Society* 19:5/6, 91–105.

Law, J. and A. Mol 2001. "Situating Technoscience. An Inquiry into Spatialities". *Environment and Planning D* 19, 609–621.

Law, J. 1986. "On the Methods of Long Distance Control. Vessels, Navigation and the Portuguese Route to India". In: ibid. (ed.), *Power, Action and Belief: A New Sociology of Knowledge?* London: Routledge, 234–263.

Lovell, H. 2017. "Policy Failure Mobilities". *Progress in Human Geography* 43:1, 1–18.

Matsumoto, N. 2006. "Analysis of Policy Processes to Introduce Bus Rapid Transit Systems in Asian Cities from the Perspective of Lesson-Drawing. Cases of Jakarta, Seoul, Beijing". Institute of Global Environmental Strategies, Japan.

McCann, E. and K. Ward 2013. "A Multi-Disciplinary Approach to Policy Transfer Research. Geographies, Assemblages, Mobilities and Mutations". *Policy Studies* 34:1, 2–18.

McCann, E. and K. Ward 2011. "Urban Assemblages. Territories, Relations, Practices, and Power. Introduction". In: ibid. (eds.), *Mobile Urbanism. City and Policymaking in the Global Age.* Minneapolis: University of Minnesota Press, xiii–xxxv.

Merry, S. 2006. "Translational Human Rights and Local Activism. Mapping the Middle". *American Anthropologist* 108:1, 38–51.

Mngodo, E. 2016. "Living a Fast Life in a Slow City Inside the Blue Bus". Accessed at *The Citizen Tanzania* (http://www.thecitizen.co.tz/magazine/soundliving/Living-a-fast-life-in-a-slow-city-inside-the-blue-bus/1843780-3256674-32b6cy/index.html). Published 19.06.2016, retrieved 17.04.2020.

Mobereola, D. 2009. "Lagos Bus Rapid Transit. Africa's First Bus Rapid Transit Scheme". SSATP Discussion Paper No. 9; Urban Transport Series.

Mol, A. 2002. *The Body Multiple. Ontology of Medical Practice.* Durham, NC: Duke University Press.

Montero, S. 2016. "Study Tours and Inter-City Policy Learning. Mobilizing Bogotá's Transportation Policies in Guadalajara". *Environment and Planning A* 49:2, 1–19.

Moore, S. 2013. "What's Wrong with Best Practice? Questioning the Typification of New Urbanism". *Urban Studies* 50:11, 2371–2387.

Morgan, M. 2011. "Travelling Facts". In: P. Howlett and M. Morgan (eds.), *How Well Do Facts Travel? The Dissemination of Reliable Knowledge.* Cambridge: Cambridge University Press, 3–39.

Mukhtarov, F. 2014. "Rethinking the Travel of Ideas. Policy Translation in the Water Sector". *Policy & Politics* 42:1, 71–88.

Muñoz, J. and A. Gschwender. 2008. "Transantiago: A Tale of Two Cities". *Research in Transportation Economics* 22, 45–53.

Ong, A. and S. Collier (eds.) 2005. *Global Assemblages. Technology, Politics, and Ethics as Anthropological Problems.* Malden, MA: Blackwell.

Orekoya, T. 2010. "The Bus Rapid Transit System of Lagos, Nigeria". A Presentation to United Nations Forum on Climate Change Mitigation, Fuel Efficiency & Sustainable Urban Transport, Seoul. March 2010.

Paget-Seekings, L. 2015. "Bus Rapid Transit as Neoliberal Contradiction". *Journal of Transport Geography* 48, 115–120.

Peck, J. and N. Theodore 2015. *Fast Policy. Experimental Statecraft at the Thresholds of Neoliberalism.* Minneapolis, London: University of Minnesota Press.

Peck, J. and N. Theodore 2010. "Mobilizing Policy. Models, Methods, and Mutations". *Geoforum* 41, 169–174.

Peck, J. 2011. "Geographies of Policy. From Transfer-Diffusion to Mobility-Mutation". *Progress in Human Geography* 35:6, 1–25.

Prince, R. 2012. "Policy Transfer, Consultants and the Geographies of Governance". *Progress in Human Geography* 36:2, 188–203.

Richter. M. 2005. "More Than a Two-Way Traffic: Analyzing, Translating, and Comparing Political Concepts from Other Cultures". *Contributions to the History of Concepts* 1, 7–19.

Rottenburg, R. 2009 (2001). *Far-Fetched Facts. A Parable of Development Aid.* Cambridge, MA: MIT Press.

Sahlin, K. and L. Wedlin 2008. "Circulating Ideas. Imitation, Translation and Editing". In: R. Greenwood, C. Oliver, R. Suddaby and K. Sahlin (eds.), *The Sage Handbook of Organizational Institutionalism.* Thousand Oaks: Sage Publications, 218–242.

Sismondo, S. 1999. "Mathematical Models, Simulations, and Their Objects". *Science in Context* 12:2, 247–260.

Star, S. and J. Griesemer 1989. "Institutional Ecology, 'Translations' and Boundary Objects: Amateurs and Professionals in Berkeley's Museum of Vertebrate Zoology 1907–39". *Social Studies of Science* 19, 387–420.

Stein, C., B. Michel, G. Glasze and R. Pütz 2017. "Learning from *Failed* Policy Mobilities: Contradictions, Resistances and Unintended Outcomes in the Transfer of 'Business Improvement Districts' to Germany". *European Urban and Regional Studies* 24:1, 35–49.

Stirrat, R. 2000. "Cultures of Consultancy". *Critique of Anthropology* 20:1, 31–46.

Strathern, M. 2005. "Robust Knowledge and Fragile Futures". In: A. Ong and S. Collier (eds.), *Global Assemblages. Technology, Politics, and Ethics as Anthropological Problems.* Malden, MA: Blackwell, 464–481.

Temenos, C. and E. McCann 2013. "Geographies of Policy Mobilities". *Geography Compass* 7:5, 344–357.

Ureta, S. 2015. *Assembling Policy. Transantiago, Human Devices, and the Dream of a World-Class Society.* Cambridge, MA: MIT Press.

US Embassy Tanzania 2009. "Three DART Officials Visit Miami's Bus Rapid Transit System". Accessed at *US Embassy Tanzania* (https://tanzania.usembassy.gov/pr_01292009.html). Published 29.01.2009, retrieved 11.11.2018.

US Green Building Council 2018. "LEED is Green Building". Accessed at *USGBC* (https://new.usgbc.org/leed). Published 2018, retrieved 17.04.2020.

Vainer, C. 2014. "Disseminating 'Best Practice'? The Coloniality of Urban Knowledge and City Models". In: S. Parnell and S. Oldfield (eds.), *The Routledge Handbook on Cities of the Global South.* Abingdon: Routledge, 48–56.

Wood, A. 2016. "Tracing Policy Movements. Methods for Studying Learning and Policy Circulation". *Environment and Planning A* 48:2, 391–406.

Wood, A. 2015. "The Politics of Policy Circulation. Unpacking the Relationship between South African and South American Cities in the Adoption of Bus Rapid Transit". *Antipode* 47:4, 1062–1079.

Wood, A. 2014a. "Learning Through Policy Tourism. Circulating Bus Rapid Transit from South America to South Africa". *Environment and Planning A* 46, 2654–2669.

Wood, A. 2014b. "Moving Policy. Global and Local Characters Circulating Bus Rapid Transit Through South African Cities". *Urban Geography* 35:8, 1238–1254.

Wright, L. and W. Hook 2007. *Bus Rapid Transit Planning Guide*. New York: ITDP.

Wright, L. 2003. "Sustainable Transport: A Sourcebook for Policy-Makers in Developing Cities. Module 3b. Bus Rapid Transit. Division 44". Environment and Infrastructure Sector Project 'Transport Policy Advice'. Eschborn: Deutsche Gesellschaft für Technische Zusammenarbeit GmbH.

4 Creating the model of the Global South

The successful dissemination of a policy model not only depends on complex translations but also on persuasive work. Such work has helped the BRT model spread, particularly within the Global South, and most recently within Africa, so that BRT has been labelled the 'model of the Global South', and DART has become known as the 'African BRT'. This chapter explores the technopolitical dimension of the global BRT model, and shows that its community of consultants, technocrats and politicians is structured by hegemonies, interdependencies and mutually beneficial relationships. First, I elaborate how technocrats attribute social, economic and political value to BRT by means of one-sided stories that favour BRT and dismiss other modes of public transport. Second, I question whether the model actually travels from South to South. Third, looking at global technocratic activities concerning DART and technopolitics within Tanzania, I analyse the assembling of DART in the context of the broader politics of BRT dissemination. Presented as the African best practice, the Tanzanian BRT system has gained a prominent position in the global BRT network.

Hegemonies and interdependencies

Since the 1960s, scholars have debated the 'dangers' of technocratic decision making for public participation and the question of whether technocrats have 'dethroned the politician' (Meynaud 1964: 13; see also Clarence 2002; Shapin 2008). Scholars have given policy actors different names: middling technocrats (Roy 2012), transfer agents (Stone 2004), travelling technocrats (Larner and Laurie 2010), intermediaries (Merry 2006), mediators (Behrends et al. 2014), policy experts (Temenos and McCann 2013) or policy mobilisers (Wood 2014, 2015a). These intermediaries produce, manage and mobilise knowledge. They can be official stakeholders from public or private sector, elected officials or voluntary consultants that act under the agenda of philanthropic values.

Consultancy and planning have always been political, competitive and contested. Linking knowledge with power helps to understand how the power dynamics of policy making shape the realisation of (transport) infrastructures (Schwanen 2016: 128; Schwelger and Powell 2008: 8–9). Global policy actors have 'extraterritorial authority' that enables them to influence developments across localities (Wood 2014; see also McCann 2011). Peck and Theodore argue that the recent acceleration of policy mobilisation 'tends to favour the kinds of technocratic strategies pushed by well-resourced multilateral agencies' (2015: xxxii). Policy actors, and more specifically consultancies, have claimed expert status for themselves in order to being able to present their knowledge and culture as objective (Stirrat 2000: 40–41; see also Boyer 2008: 39). However, technocrats do not represent universal forms of knowledge but rather ones that are highly context-specific (Healey 2012; Wilson 2006: 502). Every knowledge has its limits and contingencies.

Global planning is characterised by two intertwined processes: internationalisation and global mobilisation of best practices on the one hand, and increasing involvement of private sector actors in public planning projects on the other hand. The expanding role of global consultants has been dubbed the 'rise of a new global consultocracy' (Ward 2011: 76), and the increasing number of translocal consultancy organisations contributes to the trend of new and expanded policy networks. With its heterogeneous group of global consultants, the technocratic network has gained in complexity, becoming a global phenomenon. Consultants might be from NGOs or private companies, following different agendas and ways of persuasion. This heterogeneity may well involve competition and differing interests, and lead to power-laden processes of knowledge circulation and advocacy. Hence, consultancies can control and manipulate knowledge (Prince 2012: 198; Wood 2014: 1241) – processes that are rather opaque. In addition, there has been little research to date on how those consultants actually operate in contexts that involve development cooperation. The circulation of policies in so-called North–South collaborations may involve quite different dynamics and power relations than in more thoroughly researched cities like Vancouver, Barcelona or Frankfurt.

Introducing BRT into political debates and programmes, BRT consultants become technocrats who do not act as neutral advisors but as proponents of the model. In the process, they propagate a one-sided image of BRT, which downplays the challenges involved in planning, implementing and operating it. Proponents primarily define the BRT model by technical parameters which embody social and economic values that raise various expectations. The transport model promises not only more comfort and economic growth but also a new age of wellbeing and democracy. Proponents do not hesitate to use images, creative comparisons and metaphors to convince their audience about BRT. Even though the consultocracy of global BRT is not characterised by a heterogeneity of mobilisers, the question of power distribution is even more striking because this area of expertise underlies a strong bias and an imbalance of power: while most BRT systems emerge in countries that rely on external funding, BRT consultants are mainly attached to organisations based in North American or European countries. Scholars tend to idealise democratic decision making by contrasting it with undemocratic consultocracy (Prince 2016: 424). But as Callon et al. (2009) argue, so-called democratic processes are often not nearly as democratic as they might seem (see Chapter 5). For instance, regarding BRT in Chile, Höhnke (2012a) and Ureta (2015) describe the introduction of *Transantiago* as a technocratic and non-democratic process, because decision makers were mainly from the Ministry of Transport who took over core aspects from *Transmilenio*, such as the system's self-financing through a public-private partnership (PPP). Being primarily equipped with economic and technical knowledge, the planners focused on the financial efficiency of the system, neglecting the project's social dimensions.

BRT-cracy

Science – and therefore the technical – is 'politics by other means'; a political order is also a technoscientific order (Latour 1983: 168). Infrastructure is 'infused' with social meaning (Howe et al. 2015: 2), and is 'thought to embody solutions to structural problems' (Barry 2015). Reflecting this, Science and Technology Studies have emphasised the mutual constitution of science, technology, politics and society (see Latour 1996; Mitchell 2002; Ureta 2015). Ureta claims that the 'production of infrastructural policies never means only dealing with material or technical issues, but also enacting particular forms of political power' (2015: 4). Political actors negotiate the implementation as well as the de- and reconstruction of these orders. Technopolitics are forms of power produced through technical systems and strategic political practices, the 'hybrid forms of power embedded in technological artifacts, systems, and practices' (Hecht 2011: 3).

By linking knowledge and expertise to political power in distinctive and diverse forms, so-called experts are key technopolitical actors. Technocrats are (self)proclaimed technical experts that produce (allegedly objective) knowledge, labelled as expertise (Edwards and Hecht 2010; Harvey and Knox 2015). Latour describes the job of technocrats as 'translation-betrayal' because they translate doubts and fears into 'near-certainties' (1996: 159–160), and Björkman and Harris emphasise that the central role of engineers in producing and negotiating the everyday life is essential for understanding infrastructural processes (2018: 244). When infrastructures and technological innovations become politicised, the politics of infrastructure and the infrastructure of politics become one and the same. Prince emphasises the twofold nature of technocracy and its particular role in policy mobilities processes:

> [...] technocracy is a material assemblage comprised of particular experts and their various materials [...] technocracy exists as a particular form of rule constituted by a certain disembodied, universal rationality

that is simultaneously the product of the work of all the individual experts *and* the principle that guides their work.

(Prince 2016: 421, emphasis in original)

Thus, the entanglement of technologies with expertise and narratives can become multi-dimensional forms of politics that have concrete effects and material outcomes (Edwards and Hecht 2010: 620; Hecht 2011: 2–3).

Globally circulating models are always marked by the mutual influence of policies, policy actors and places. During their mobilisation and translation, policy models mutate, but the humans involved are also reshaped by new knowledge and experiences. Mutation can occur subtly, as when policy actors slightly change their views, or more obviously, as when they acquire new positions and identities (Larner and Laurie 2010: 219; McCann and Ward 2013: 9). Because they are necessarily transformed by the processes in which they are engaged, policy actors are neither neutral nor objective in their decision making. Despite their different disciplinary and national backgrounds, consultants and their teams share a particular rationale by which they try to 'format' the world according to universal principles and imperial visions (Peck and Theodore 2010: 173). Consultants might assume their own background environments are superior and try to influence foreign governments to implement transport systems that resemble those of their hometowns. Context-specific knowledges and practices risk being ignored, downplayed or dominated by other forms of knowledge.

The BRT market is no exception. It is dominated by international consultants who enable the global dissemination of BRT, and the World Bank and ITDP (Institute for Transportation and Development Policy) recommend or even demand including international consultants in a BRT planning process (BN 10/2015; JK 05/2016; see also Wright and Hook 2007: 96). Thus, the contributions of local planners are rarely recognised (AE 05/2016; WG 10/2015), further cementing the central role of consultants in planning processes (Grijzen 2010). In this field, expertise and services are offered not only by profit-driven enterprises but also by donation-based, charitable organisations. This BRT community, whose rule, activities and power I have labelled 'BRT-cracy', has propagated a number of normative narratives designed to persuade politicians. Situated between globally circulating models and local appropriations, consultants can impose ideas according to their own values and interests (Merry 2006: 48). But precisely because their decisions are based on their own interests, normativities and values, the rationality and motivations of global consultants and other global policy actors need to be questioned. Due to a 'strong-willed politics of persuasion' (Temenos and McCann 2012), local politicians and technocrats tend to follow the advice of global consultants.

In contrast to Bowker's claim that infrastructures 'do not have plotlines or heroic figures' (2015), BRT has both: a story of success as well as prestigious leaders and best practices. Led by the heroic figure Enrique Peñalosa, the BRT model in general, and each implemented BRT system in particular, has its story to tell. Technocrats constantly embellish and update the global BRT plotline. This storytelling (Czarniawska 2004) is so pervasive that scholars describe BRT as being 'pushed from the outside by charismatic proponents' (Wood 2014: 1251), and question why BRT is being promoted so aggressively (Rizzo and Dave 2018: 2). A triad of fame – the heroic figure Peñalosa, the global best practice *Transmilenio* and the city of Bogotá – plays a central role in BRT's global story of success.[1] One-sided stories focus on *Transmilenio*'s achievements and disguise critical points. One of DART's principal consultants told me: 'Bogotá is a good example for everything' (LL 11/2016). Transport researchers identify *Transmilenio*'s main strengths as the incorporation of the existing public transport operators, its high passenger capacity and its ability to operate without direct public subsidies (Mirailles 2012: 9; Pineda 2010: 124). However, the main reason that *Transmilenio* is considered the global best practice is the activism of Peñalosa and his network. The strong rhetoric of Peñalosa swallows all criticism so that stories about 'the miracle of Bogotá' (Berney 2017) and 'la receta Transmilenio' (Flores 2011) prevail. As a World Bank retiree told me, the global BRT community has put heavy emphasis on *Transmilenio* after its inauguration in 2000 in order to foster the global dissemination of the BRT model: 'Bogotá was the champion of BRT; it became a center of learning for everywhere' (DN 03/2015). Since then, numerous cities like Dar es Salaam, Guadalajara, Santiago de Chile, Johannesburg and Jakarta applied Bogotá's BRT model so that *Transmilenio* has become the most popular BRT system worldwide. More than 10,000 decision makers made study tours to Bogotá (Wood 2015b: 1062–1069), and Peñalosa went on a tour in 2003, where he 'inspired change across Africa' that led to a turning point in Cape Town and other African cities including Dar es Salaam (ITDP 2003; see also Wood 2014: 1244–1245).

This triad shows to what extent global policy circulation can depend on individuals, backed and morally supported by a community. ITDP and the World Bank benefit from this renowned triad likewise. BRT is a promising investment for development banks, which finance the vast majority of BRT projects – and thereby ITDP's projects – in the Global South. According to Rizzo, the global BRT network comprises various individual and collective actors, joined together in a tightly knit web of institutions he dubs 'the BRT Evangelical Society' (2017: 148ff.). This network is small: besides Peñalosa and ITDP at the core and the World Bank as main financier of BRT projects, only a handful of other individuals and organisations comprise the global BRT community, several of them collaborating in collecting data, standardising and advertising BRT.[2] Global technocrats are involved both in the initial planning and implementation phases of technopolitical projects and also in later stages of operation. The CEO of the DART Agency shared with me his experience: 'With such a project, you kind of start from scratch. There are so many things involved. [...] You need a wide range of experts' (GV 09/2016). By 'experts' he meant international experts.

The second global BRT protagonist alongside Peñalosa, ITDP, is a ubiquitous presence in BRT projects. Despite being a comparatively small organisation, ITDP has succeeded in taking the central position in the global BRT community (ITDP 2020). The NGO functions as a consultancy that provides political and technical guidance. Besides its strong global connectivity, its success mainly rests on two strategic characteristics. First, ITDP has developed regional networks in which locally staffed regional offices liaise directly with the US head office. Second, ITDP promotes a strong work ethic and the staff are highly committed to and profoundly convinced by the NGO's values. The director of the ITDP office in Nairobi described his decision to leave the private sector to join ITDP as 'moving to the good side' (PH 03/2015). He wholeheartedly believes in BRT and expects his staff to show high personal commitment to this work. ITDP refers to its employees worldwide as a 'family' and lists a 'strong commitment to advancing ITDP's mission' as key qualification of its employees. Moreover, ITDP benefits from its tight relationship with Peñalosa and collaboration with other global BRT technocrats (ITDP 2015b; Wood 2014: 1249). Bringing together politicians with representatives of development cooperation agencies and bus manufacturers, the NGO's board of directors and technical committee annually award cities with the *Sustainable Transport Award* (STA) and evaluate BRT systems for the *BRT Standard* (see Chapter 3). The NGO also maintains its superior position in the global BRT community by collaborating closely with local consultants and planners, and it proficiently connects heroic figures with intermediaries and consultants who have context-specific knowledge. Already in the first edition of the *BRT Planning Guide*, ITDP's consultants emphasise the benefits for city governments in collaborating with international experts, taking the case of Peñalosa visiting Dar es Salaam in 2003:

> A local team working in conjunction with experienced international professionals can ideally result in a combination of world best practice and local context. [...] Dar es Salaam has successfully combined a full local team with international consultants.
>
> (Wright and Hook 2007: 96)

In addition, ITDP publishes the central BRT planning and mobilisation tools. ITDP has had projects in over a hundred cities, mainly in in the US, Latin America, East and South-East Asia and Africa. Under the umbrella of sustainable cities, ITDP's consultants nowadays focus on BRT, non-motorised transport (NMT) and transit-oriented development (ITDP 2015a). ITDP is constantly striving to keep its top place in the global BRT community by pursuing a dual role: as an NGO supporting sustainable cities on the one hand, and as a consultancy specialising in BRT on the other. In this dual role, the organisation provides basic information on BRT to local authorities, connects local and global actors, assists with initial designs and the allocation of funding. ITDP tends to be less involved in the later stages of planning and operations, but it remains on stand-by for ad-hoc advice.

The DART project needed international expertise for long-term training of drivers and mechanics from the Chinese bus manufacturer as well as training on the use of the intelligent transport system (ITS) from the Austrian software supplier. These after-sales services not only reveal the dependency on knowledge from elsewhere but also the continuing expenses that the project demands from the Tanzanian government, service operator and investors. An employee of the operator UDA-RT (Usafiri Dar es Salaam Rapid Transit) urged the World Bank to extend funding beyond the construction to the operational phase since the government will not be able to subsidise DART's operations (AH 05/2016).[3] Being a borrower over decades, the Tanzanian government has to fulfil the conditions that the World Bank attached to the loan. In this case, the World Bank requested to include one international company under an international PPP arrangement. However, after the loan had been disbursed, Tanzanian companies tried to circumvent this international collaboration so that the World Bank no longer had a direct means to enforce this condition. Having successfully blocked the international tender for a second bus operator, UDA-RT started operations as an interim service in May 2016. Subsequently, the World Bank showed interest in financing Phase 3 and Phase 4 of DART (BN 10/2015; HM 09/2016), and thus was able to attach the condition of an international PPP framework to the funding of Phase 1.[4] The Tanzanian government would only receive the official confirmation of the new loan when an international operator for Phase 1 is guaranteed, as the CEO of the DART Agency explained: 'They are looking more closely now, before accommodating the loan, they look at the performance of Phase 1. [...] Because no one wants to invest in a project that is not doing well'[5] (GV 05/2016). The World Bank not only insisted on this condition because international PPPs are one of the bank's principles, but also because the bank was interested that DART would operate according to international standards enforced by an external entity – as the quality of operations could also fall back to the bank's reputation.

This case shows that infrastructural projects are highly political and that international organisations are not the neutral, altruistic partners they claim to be. Donors in particular exert power through the conditions they attach to project implementations. Several Tanzanian government officials told me they would prefer to make deals with banks that do not attach conditions to their funding – for instance, Chinese banks offer deals that are free of interference on a normative level. However, World Bank officials like James Kombe deny that they exert influence on government decisions: 'the World Bank has no ideology' (JK 05/2016), and projects must only align with the bank's goals such as poverty reduction to be funded: 'We don't support everything' (JK 09/2016). World Bank officials told me that the bank does not have a fixed position on DART's changing service provider model; as long as general principles were followed and the main project objective is not affected, details can be changed during the process. For that reason, the World Bank agreed to the solely Tanzanian interim service, but not to changing the international PPP to an exclusively national PPP (see Chapter 5).

Favouring BRT

Since ITDP and BRT are deeply interwoven, their success in terms of global proliferation and growth is mutually dependent. The historical development and spatial dissemination of ITDP and BRT parallel each other closely. When *Transmilenio* received international attention and the label of best practice in the early 2000s, ITDP's staff size and annual budget were also growing (ITDP 2015b). ITDP field offices popped up wherever BRT systems were developed; conversely, cities invested in BRT precisely because of ITDP's presence. In practice, ITDP favours BRT not only over private cars but also over other modes of public transport, although it does not officially acknowledge this privileging because that would challenge its objectivity as a consultancy (c.f. ITDP 2018m). In its projects, publication and presentations, IDTP equates BRT with public transport and high-quality transport systems that 'make cities more livable, inclusive, accessible and sustainable' (ITDP 2018a); rail-based systems do not appear. For instance, ITDP's conference presentations on mass rapid transit options for East African cities and on how to improve urban mobility only mentioned BRT, implying that it was the only solution worth considering for East African cities. At a transport conference that I attended in Nairobi in November 2016, the convenors and participants criticised ITDP for disregarding light rail and other MRT options. City officials argued that ITDP had delayed the decision on their future main mode of public transport by rendering rail absent.

ITDP would not have this uncontested position without being permanently present through its various BRT mobilisation tools (see Chapter 3) and without the support of other actors of the BRT community. Transport planners from universities, development cooperation agencies and governments esteem ITDP's work, including the quality of its publications, its grassroots activism and its broad knowledge transfer. The former CEO of the DART Agency described ITDP's engagement in Dar as a 'blessing' (YK 10/2015), and a transport researcher characterised ITDP as an 'honest broker in forging agreements across political factions' (TG 12/2016). During decision making processes, the NGO emphasises the autonomy of local governments and downplays its own impact. Nevertheless, politicians and planners question ITDP's BRT favouritism. As a Tanzanian transport planner elaborated: 'They [ITDP] evoke quite strong feelings. [...] One accusation is that they're just pushing their own ideas without the due respect for the actual needs of the society or country' (NT 10/2015).

This 'pushing' is rooted in the NGO's legal and financial status because it has to follow the agendas of transport financiers. Since ITDP has no budget of its own, it heavily depends on funders and hence needs to align its advocacy work to global trends of development cooperation agencies and banks, political foundations and funds. ITDP uses its trustworthiness as an NGO and its longstanding activities within global transport networks as leverage to influence global trends. Consequently, ITDP emphasises its role as an NGO that fits into financing schemes of the World Bank, the United Nations Environment Programme (UNEP), the German Agency for International Cooperation (GIZ) and others, and emphasises its consultancy work on narratives of social equality, sustainable transport and 'livable cities' (ITDP 2018a). Since BRT fits perfectly into these narratives and development ideals, this public transport model helps ITDP with fundraising on the one hand, and, on the other hand, BRT to prevail as a practicable, lucrative and worthwhile solution for cities of the Global South.

Hence, planning and mobilising BRT is political, emotional and normative. For members of the global BRT community BRT is more than transport (ITDP 2018d), as they attribute meaningful terms like development, democracy, dignity, liveability, equity and equality to the transport model. By contrasting private and public transport, both ITDP and Peñalosa turn urban mobility into a question of who rules urban (road) space: whereas BRT and NMT stand for the people and democracy, private motorised transport is associated with inequality and social fragmentation (ITDP 2018l; Peñalosa 2013). Peñalosa argues that more equal, just and inclusive cities can be created through transport innovations like BRT, whose lanes are 'symbols of respect, equality and human dignity' (Peñalosa in Project for Public Spaces 2008). According to the Colombian politician, BRT is 'democracy in action' because the system gives more priority to humans than to cars:

The first article in every constitution states that all citizens are equal before the law. That is not just poetry. It's a very powerful principle. For example, if that is true, a bus with eighty passengers has a right to eighty times more road space than a car with one. We have been so used to inequality, sometimes that's before our noses and we do not see it. Less than hundred years ago, women could not vote, and it seemed normal, in the same way that it seems normal today to see a bus in traffic. In fact, when I became mayor, applying that democratic principle that public good prevails over private interest, that a bus with a hundred people has a right to hundred times more road space than a car. We implemented a Mass Transit System, based on buses on exclusive lanes. We called it *Transmilenio* in order to make buses sexier. And one thing is that it is also a very beautiful democratic symbol, because as buses zoom by, expensive cars stuck in traffic. It clearly is almost a picture of democracy at work.

(Peñalosa 2013)

Accordingly, *Transmilenio* is the ultimate egalitarian transport solution because the buses physically unite people from diverse income and status groups; in this space, a Vice-President and a doorman would meet as equals (Ardila-Gómez 2004: 332). Contrasting rich and poor, private and public transport, Peñalosa further brings a new twist to understanding of development: 'A developed country is not a place where the poor have cars. It's where the rich use public transportation' (Peñalosa n.d.). This quote might be his most famous and was cited to me by transport consultants and politicians in Dar es Salaam several times. Moreover, as Wood's and Rizzo's metaphors show, BRT's politics of persuasion is about believing in the transport model: Peñalosa is the 'messiah', the *BRT Planning Guide* is the 'bible' and BRT promoters are the 'Evangelical Society' (Rizzo 2017: 26; Wood 2016: 399).

Opposing BRT

Both the general public as well as politicians used to perceive buses as an inefficient, unreliable and polluting mode of public transport that was primarily used by the poor (Hensher 2007: 99; Lagendijk and Boertjes 2012: 295) – until Peñalosa and ITDP succeeded in changing the image of 'old' buses to 'sexy' buses (ITDP 2018d; Peñalosa 2013). Peñalosa has challenged the 'supremacy of metro systems' (Mi30ailles 2012: 8) by demystifying the belief that people 'naturally' prefer trains to buses (Jaffe 2015). Nevertheless, cities searching for low-budget MRT solutions continue to show interest in LRT (light rail transit) and other rail-based options, as I experienced at the transport conference in Nairobi:

> A delegation from Addis Ababa, which has just recently inaugurated one of the first LRT systems in Sub-Saharan Africa, is participating in the conference. Delegates from several East African cities address various question on BRT and light rail transit (LRT) in order to find out about the best solution for their urban transport networks: which system has the higher passenger capacity at lower costs? ITDP employees stress the advantages of BRT, especially that the performance of BRT can be even better than that of LRT, at significantly lower construction costs. This argument is effective – ITDP convinces the conference participants that BRT was the better choice. Even the Ethiopian team decides on BRT for future transport improvements in Addis Ababa at the end of the conference. Ethiopian transport officials state that they could have built six BRT lines for the cost of their two LRT lines. One of them tweets: 'The LRT was political but now we have realised our mistake and making it right' (ITDP Africa 2016). During to the conference, the director of ITDP Africa corresponds with Ugandan transport officials in Kampala. He finally also convinces them to leave their LRT plans behind and switch to BRT. The Ugandan

Ministry of Transport names its recent very impressive visit to Dar es Salaam as a decisive factor.

(November 2016)

Hence, ITDP was able to persuade African conference delegates to choose BRT, and not only because of the model's economic promise: unlike BRT, LRT lacks both global heroic figures and a clear plotline (see Bowker 2015).

Because they are so convinced of the superiority of their model, protagonists of the global BRT community accuse rail-focused companies and governments of using unethical, self-interested methods to persuade policy actors. The former CEO of ITDP and co-author of the first *BRT Planning Guide* writes in another BRT publication:

Naturally, light rail and metro interests are threatened by the proliferation of cheaper, more flexible BRT systems. Rail interests in the United States, and particularly companies from France, Germany, and Japan, are financially threatened by the rapid proliferation of BRT. Japan's technical cooperation agency, JICA, and to a lesser extent the French and German governments, have been active around the world promoting their rail companies by disseminating misinformation about the limitations of BRT systems. They finance feasibility studies that tend to exaggerate the projected ridership and financial feasibility of proposed light rail or metro systems.

(Hook 2009: 30)

This accusation is ironic because BRT proponents do exactly the same sort of persuasive work, even if their economic self-interest is less obvious. The general tensions between ITDP, the World Bank and JICA exemplify the competition between rail and BRT in cities that depend on external funding for infrastructure projects. In Dar es Salaam, World Bank employees, who themselves deny acting according to their own interests, accuse JICA of unfairly favouring rail projects. As the publisher of Dar es Salaam's urban transport master plan, JICA has the means to enforce its own interests in the urban development. BRT proponents claim that JICA only included DART in the master plan because DART had already been so far developed that the JICA could not ignore it any longer. Even though the master plan scheduled 2030 as its target year, JICA already started to modify the plan in 2016. The director of JICA's development project department argued that this review was necessary because BRT could not accommodate Dar es Salaam's growing transport demand, and thus the city needed to change its plans from BRT to rail-based MRT (HK 05/2016). To support his argument, he claimed that BRT's maximum capacity was 15,000 pphpd (passengers per hour per direction) whereas the plan predicted a demand of 20,000 to 40,000 pphpd on the planned BRT corridors (World Bank 2017). JICA estimated a much lower carrying capacity for BRT than ITDP, which calculates a maximum capacity of 45,000 pphpd (ITDP 2017c) – where exactly the truth lies, no one can clearly say yet. A transport consultant at the World Bank however told me that he does not blame JICA for favouring rail because JICA is under pressure to give something back to its taxpayers (JK 10/2015). In fact, most of JICA's investment goes back to Japan; just as a major share of the World Bank loan for DART went to European companies.[6]

From South to South?

Due to the narrative of high capacity at low cost, BRT is labelled as a model for the Global South. BRT technocrats justify this argument on two main grounds. First, cities of the Global South have similar structures, including prevailing minibus systems, rapidly increasing mobility needs, serious congestion, fast urban growth and urban sprawl. Second, these cities have only limited financial resources for improving their transport conditions and populations who depend on public transport cannot afford high fares. Thus, metro systems are considered appropriate for the Global North while BRT is 'for the poor' (Cervero 2013). The engineer Luc Lavoie from the Brazilian transport consultancy that did the initial design of DART, summed up this argument:

> With BRT, you can do a network in 10 years. That's what Bogotá have done. [...] You can really change the city in very short time. If you do metro, LRT, it's very expensive. It costs too much. And so, you depend on technology and everything. So, in my opinion, all the developing countries need to do BRT.
>
> (LL 11/2016)

More generally, scholars and consultants assess that 'models and solutions used for cities in the developed countries may not be applicable to cities of developing countries' (Msigwa 2013: 8), where both the limited financial resources of city governments as well as the low average income of those who depend on public transport support the argument for (comparably) low-cost BRT operations. However, such assumptions are based on a homogeneous image of cities of the Global South, even though conditions vary widely between cities. For instance, in the two East African cities Dar es Salaam and Nairobi, legal forms of minibus networks and general policy frameworks differ significantly.

Not only is BRT argued to be a transport mode appropriate for the Global South, it has supposedly been developed within the South. Global consultants describe the planning and implementation process of Dar es Salaam's BRT as a transfer from the South to the South:

> For me, it's a South–South transfer. Because really, BRT was developed in Latin America. [...] And I'm really happy that this is a South–South

knowledge transfer. Actually, when we want to know, the knowledge of running BRT, we have to go to Latin America, not to Europe.

(HM 03/2018)

The initial inspiration for DART came from Bogotá, a city that might be referred to as in the Global South. But does it justify the depiction South–South; and which actors do in fact stand behind circulating BRT?

The dominant BRT narrative of Peñalosa and ITDP leaves out the US and Europe; whose BRT systems have also impacted the global BRT model. However, the reality is that most of the leading BRT individuals and or-ganisations have a US or European background, a fact that shapes how the BRT model circulates globally and is materialised in different places. Despite having its headquarters in the US, ITDP puts a lot of effort into appearing as a global NGO that is primarily based in the Global South. The circulation of information and expertise on BRT within the Global South is actually quite limited. Whereby Bogotá's *Transmilenio* appears frequently, some cities are hardly mentioned at all. In Dar es Salaam, people wondered why there had been so little exchange within Africa – like between Tanzania and South Africa – during the planning phase of DART (DN 03/2015; RP 05/2016). One reason was that ITDP had not been involved in developing BRT in South Africa, and lacks relationships to South African institutions and operators. Powerful actors decide which BRT systems receive attention and fame – and which do not. Wood thus concludes that ITDP is 'a northern entity with a southern eye' (2015: 1074).

Mawdsley describes the consolidation of South–South development as a 'rupture in the North–South axis' (2017: 108), in which Southern actors also become providers of expertise. Organisations like the World Bank increas-ingly extend their so-called South–South facilities accordingly, in which for instance Vietnam and Brazil serve as 'provider countries' of expertise on ur-ban transport programmes for 'recipient countries' like Indonesia and Tan-zania (BN 10/2015). I experienced these supposedly Southern relations at the conference in Nairobi, which was organised by GIZ and UN Habitat with the goal of developing a regional strategy for East Africa and supporting South–South exchange. Representatives of DART presented the Tanzanian BRT system as best practice solution for transport problems in East Africa. The asymmetry in the development sector could be plainly seen: between those who create knowledge, develop programmes and take decisions on the one hand, and those who follow the advice and fulfil the conditions on the other hand; between those who decide which projects merit being financed, and those who need external funding in order to conduct infrastructure im-provements. The conference culminated with the final words of a leading GIZ employee who secured further guidance and funding for the partici-pating cities, earning the gratitude of the East African participants. Hence, South–South development is often not as Southern as it seems at the first sight.

Labelling BRT as a transport solution that is not only *for*, but also *of* the Global South helps the BRT model to proliferate. Behrens and Ferro even attribute a South–North transfer to BRT:

> Bus transportation arguably represents one of the few policy sectors in which innovation and technological development has flowed stronger from the 'global south' to the 'global north', than the other way around.
>
> (2016: 203–204)

However, looking at the physical infrastructure and the technical expertise DART inhabits, the picture becomes more complex. In order to be able to operate DART, its stakeholders need to receive knowledge from various places. Whereas knowledge of bus maintenance, driving and scheduling came from Chinese and Indian partners, knowledge of PPP arrangements and operations with an ITS came from European and US consultants. Therefore, translating and materialising BRT is a complex process that involves actors, expertise and materials from various places.

One-sided stories

Many actors have an interest in DART's success, ITDP most of all. The NGO needs successful projects so that it can attract further funding and develop further BRT projects; it thus supports DART in order to gain support for itself. ITDP uses its DART-inspired vision for sustainable transport in African urban centres in its funding drives:

> In Africa, ambitious cities like Dar es Salaam, Tanzania are leading the charge with major commitments to affordable, sustainable mass transit. With extensive support from ITDP, Dar es Salaam's groundbreaking new bus rapid transit (BRT) system—the first true BRT in East Africa—is setting the bar for sustainable African cities. [...] Will you help cities like Dar es Salaam continue to transform the ways they move and grow? Become a $10 ITDP Monthly Giver today.
>
> (ITDP 2017d)

In the process, however, ITDP glosses over DART's many technopolitical challenges and conflicts by telling one-sided stories. Sharing successful attributes and downplaying difficulties by 'selling' instead of 'telling' their stories to city officials is a common strategy of policy mobilisers (Wood 2014: 1241–1246).

Is ITDP therefore not an 'honest broker' (TG 12/2016)? The organisation's strategy consists of not only disseminating BRT as a successful transport model but also disseminating the success of specific BRT projects, allegedly proved by the NGO's strong narratives and self-referential tools like the *BRT Standard*. ITDP's stories seem to verify themselves: telling a successful story of BRT or DART enhances the probability of success. ITDP predicts the future of BRT, an act that is already the first step towards that future. Thus, at the conference in November 2016, ITDP did not inform the participants about the difficulties and reasons for the delays of DART's inauguration. Employees from ITDP's Africa Office only talked about DART's short travel time, its high acceptance by the citizens and better comfort compared to daladala. By contrast, the CEO of the DART Agency openly spoke about DART's difficulties and gave valuable insights.

Another example of one-sided BRT storytelling is the silence in the global BRT community regarding the controversial planning process of the *Transantiago* transport project. The project led to a collapse of public transport in Santiago de Chile, public demonstrations and a crisis for the national government that extended far beyond the transport sector (Höhnke 2012a, 2012b; Ureta 2015: 138). As Ureta (2015) argues, *Transantiago*, far from delivering the promised world-class system, was a prime example of policy failure and non-democratic decision making. However, this technopolitical crisis is largely absent in ITDP's descriptions of *Transantiago*. Even though the 2017 version of the *BRT Planning Guide* does point to some of its problems, such as the incomplete infrastructure and the system's insufficient service and communication, it downplays the gravity of the technopolitical controversies, describing them merely as 'initial confusion and operational problems' (ITDP 2017c: 334, 521). Moreover, the NGO scored two corridors of *Transantiago* as *Bronze* in 2014 and awarded Santiago de Chile with the STA (Sustainable Transport Award) in 2016 (ITDP 2017b).

ITDP publications paint a similarly rosy picture of the BRT model, publishing images of DART and other systems of high-tech, shiny and imposing infrastructure and smiling passengers. Overcrowding, operational delays or accidents are rendered invisible. Such images contrast with my own experiences gathered in Dar es Salaam and in the chat group *BRT Education Campaign*. Generally, ITDP is very receptive to different perspectives, but certain individuals are so profoundly convinced of BRT that they deny any fundamental critique. Even though the global ITDP team does not position itself as a clear BRT proponent, the NGO keeps a coherent image on the outside by using storytelling as a 'sense-making process' (Boje 1991; see also Cameron 2012). ITDP produces, sells and circulates stories that others consume, translate and only sometimes contradict. These practices of one-sided storytelling recalls Latour's notion of Janus' four dictums and the contrary perspectives of ready-made science and science in the making: 'Once the machine works people will be convinced' versus: 'The machine will work when all the relevant people are convinced' (1987: 13). The two perspectives need to consider each other: only when all actors of the assemblage harmonise will DART work according to the norms inhabited in stories and plans. Neither a purely technical solution nor strong storytelling and politics of persuasion will make DART become an actual success.

ITDP's one-sided and partial stories are highly performative as they dominate how DART is viewed globally and convince policy actors about the (presumed) lack of alternatives to BRT. On an ITDP webinar, the chief programme director spoke highly of DART's vision:

> It starts with setting the vision: Being clear about what success is gonna look like for you. What do you want to achieve with this? It's about having your goals and objectives for this model apparent. [...] In Dar es Salaam, they were very clear about what their objectives were. And they were clear about their values.
>
> (ITDP 2018b)

The director did not mention to what extent DART had failed to fulfil its objectives and values because this would have cast a cloud over the so-called African best practice. The presentation further gave policy makers the impression that the aim and the vision of a planned BRT system would materialise without mutating, and that its success is guaranteed. But even if positive stories slightly strengthen BRT's position, they do not guarantee actual success – neither of a given BRT project nor of ITDP's work in general.

The African BRT

ITDP discovered the potential of the African market for BRT systems, resulting from pressing transport needs and the willingness of donors to fund BRT projects in the region. ITDP wants to continue to grow and expand its activities on the continent. After opening the regional office in Nairobi in 2015, ITDP plans to establish an office in West Africa in the near future. The Africa Office in Nairobi exemplifies ITDP's strategy of combining contextually bound and unbound knowledge. While the director brings experience from the US and India, the local team contributes context-specific knowledge on national legislation, institutional structures and local languages, as well as personal and professional networks with Kenyan politicians and institutions. The team in Nairobi underlies a clear hierarchy because the director, showing clear leadership, is closely linked with the head office in the US. He allocates responsibilities, decides which projects to work on, and has the final say in every presentation and publication. As the name already indicates, the regional office is not only responsible for transport projects in Kenya, but also for (mainly BRT) projects in other African regions (ITDP 2018d). Regional offices elsewhere are named after the respective city or country, but not the continent. The Africa Office is well located in Nairobi, the heart of NGOs and international organisations in Sub-Saharan Africa, so that ITDP can connect well to other projects and financiers. The NGO not only promotes BRT as 'the new face of public transport in Africa' (see Chapter 3), but also advertises Dar es Salaam's transportation improvements as leading the 'breakthrough for African Cities' (see Figure 4.1).

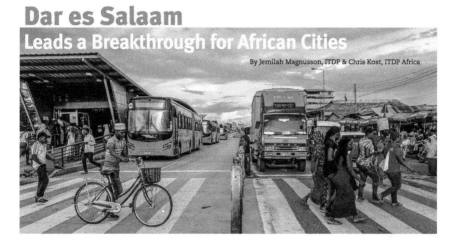

Figure 4.1 According to ITDP, DART facilitates a 'breakthrough' for African cities (ITDP 2018e).

According to the former CEO of the DART Agency, ITDP told the Tanzanian authorities in 2003: '"We're creating Bogotá for Africa"' (YK 10/2015). All leading figures of DART tell the same story of BRT's arrival in Dar es Salaam: With funding from the World Bank, UNEP and the Global Environment Facility (GEF), ITDP and Peñalosa had the objective of bringing BRT to Africa, with Dar es Salaam as a pilot city, as the chief technical advisor remembers:

> [Peñalosa] came here, he met with the mayor of Dar es Salaam and he told him about. And he invited him, the mayor, who was Sykes. And he went to Bogotá. And he saw it. And then, he convinced him that this is, that's what we need in Dar es Salaam. That's the way it started.
>
> (HM 03/2015)

The famous triad of Peñalosa, *Transmilenio* and Bogotá has been DART's constant companion. In 2003, the former CEO of the DART Agency attended Peñalosa's BRT workshop in Dar es Salaam and remembers: 'Peñalosa was very famous. And, very moving statements. Then, everything from Bogotá was considered to be ideal' (YK 10/2015). Planners copied *Transmilenio*'s physical design for DART; especially the stations are rated as a 'replica' (DS/MH 09/2015; YK 10/2015). However, the infrastructure does not look that much alike. DART has a different colour design, stations started to degenerate even before the inauguration of the system, and *Transmilenio*-inspired components like automatically sliding doors and visual displays have not materialised.

Initially, Dar es Salaam city officials had the quite unpretentious goal of reducing congestion. Yet, Peñalosa advised them to also focus on the attractiveness of the city, i.e. to utilise DART to attract investments for Dar es Salaam and to facilitate a broader urban regeneration (YK 10/2015). By that time, *Transmilenio* had just started to operate its first line and Peñalosa was on ITDP's board of directors. The 'tightly knit web' (Rizzo 2017) promised Tanzanian officials that Dar es Salaam would go through a similar urban transformation as Bogotá had. Peñalosa and ITDP presented a version of *Transmilenio* to the Tanzanian authorities that was filled with grand visions and elided the controversial parts, as the former CEO of the DART Agency remembers:

> In 2007, I was participating in one of the trainings in Sweden on urban transport. We had a representative from Bogotá. He said: 'Oh, you know, if you come to Bogotá, there's too much of sugar-coating. Not everything what they tell us is real.' [...] If they could have told us the bad things, then it was not easier for us to move forward. It could not be encouraging. Then sometimes, it's good to put the sugar so that we move. It has inspired us to move forward.
>
> (YK 10/2015)

The former CEO felt motivated by the sugar-coated version of *Transmilenio* and was not very fond of learning about the system's difficulties. This attitude might have led ITDP to describe Dar es Salaam as 'enthusiastic' about BRT (ITDP 2018n). But not everyone in Dar es Salaam has been enthusiastic about DART so that the Tanzanian government has been in a difficult position throughout DART's planning and implementation process, trying to internally mediate between opposing positions. Even though the official position of the Tanzanian government is to stand behind DART – in fact, no government official ever called DART into question in front of me – Rizzo identifies 'mixed motivations of the CCM' and a 'tepid commitment to BRT at the heart of government' because not everyone was expected to benefit from DART's promised win-win (2017: 147). This tepid commitment has not only led to the slow, intermittent implementation of the project, but also to a lack of political leadership behind the DART project. Nevertheless, the 'half-hearted enthusiasm' and the 'lack of a strong local engine' that not only DART, but also *Aramis* was afflicted with (Latour 1996: 85, 159–160), have led to longstanding project delays and technopolitical controversies in Dar es Salaam. Thus, maybe the Tanzanian government has not been as enthusiastic as ITDP claims?

DART persists as the interface between two Southern spaces: it is not only a site of BRT arrival, but also of BRT departure. The BRT project, referred to as the 'African guinea pig' (HM 03/2015) has become the point of reference for disseminating this transport model throughout the continent because 'Africa is the new Mecca of urban transport' (HM 03/2015). Global consultants use DART as a starting point for entering the African transport market, as transport actors in Dar es Salaam described it. Amir Hanif, the spokesperson of the service provider said: 'If all goes well, I think, BRT will be the future mode of transport in Africa. [...] All eyes are on us' (AH 10/2015). The CEO of the DART Agency also referred to the weight of expectation this pilot project brings to the city: 'We are increasingly getting more visitors. They want to open up [their own BRT system], so they come here to see how it works'[7] (GV 05/2016). Since the inauguration of DART's Phase 1 interim service operations in May 2016, transport planners, politicians and journalists from various African countries have visited DART. Some of them participated in a DART-inspired workshop on transport innovations in Dar es Salaam, organised by ITDP. The director of the ITDP Africa Office already predicted DART's learning effect for other African cities: 'One trip to Dar would solve so many problems in Lagos' (PH 11/2017). In its publications, ITDP is using every opportunity to exhibit DART an inspiration for other African cities:

> DART is more than a public transport system—it offers improvements for pedestrians and cyclists as well. With cities across the region planning their own BRT systems—Addis Ababa, Cairo, Dakar, Kampala, Kigali, and Nairobi, to name a few—the DART system offers a unique opportunity to see a high-quality BRT in the African context. [...] The expanded system will continue to inspire cities around the continent, demonstrating that investments in public transport are well worth the hard work, both economically and politically.
>
> (ITDP 2018e: 6–7)

Policy actors in Dar es Salaam are willing to collaborate in spreading the African BRT, and global organisations mobilise DART's protagonists from Tanzania to disseminate the model across the continent. The clear plotline and heroic figures of BRT's global model succeed in maintaining and (re)producing the image of BRT as a model of the Global South, with regional, supposedly 'African' specificities. ITDP needs DART in order to pursue its goal to expand the BRT market in Africa, and therefore constructs DART as the African version of BRT without determining what this 'African' is about. Just as Latour (1996: 18) describes the private rapid transit *Aramis* as a 'collective dream' of planners and politicians, BRT can be seen as the collective dream of the BRT community for the Global South, and DART as the collective dream for Africa. However, numerous policy actors in Dar es Salaam have the ambition of making DART comparable with public transport systems in the US or Europe. In December 2017, a member of the WhatsApp group commented on a picture of a nocturnally illuminated DART terminal: 'Amazing. [...] Like [in the] US. As if this was actually in Europe!!! It is nice at our place!!!'[8]

Express the success!

> A child is born inside Kimara Terminal. The mother gives her child
> the name *Dartlicious*. The transport company comments: 'Usually,
> children born on airplanes are gifted with lifelong free travel with
> the respective airline. Now, UDA-RT considers giving this present
> to Sarah and her new daughter'[9] (UDA 2017: 7). A person named
> after DART – this is definitely a success!
>
> (June 2017)

Success or failure? This binary appears frequently in the description of travelling models and mobile policies. Research partners and their means of mobilising the transport model, such as guides and webinars, generally refer to BRT and DART as a success. The *BRT Planning Guide* (2017), for instance, uses the word 'success' on at least every sixth page and illustrates it with meaningful quotes like: 'Success is a journey, not a destination' (Sweetland in ITDP 2017c: 154). By contrast, scholars argue that success and failure are intertwined and relationally constituted (Lovell 2017; Stein et al. 2017). Regarding technopolitical projects, de Laet and Mol claim that success is not an either/or matter. Technical objects that function other than planned can still be considered to be functioning successfully (2000: 227). Others stress the need to take failure into consideration when researching infrastructures or international planning (Graham and Thrift 2007: 7; Roy 2005: 156).

The notion of success has a crucial impact on the global dissemination of the transport model. Diverse actors in Dar es Salaam adopted this narrative, and the CEO of the DART Agency presented DART accordingly at the international transport summit in Dar es Salaam in September 2016:

> I think, what we have done in Tanzania is remarkable and should be
> emulated in other places, in other cities in Sub-Saharan Africa. We had
> a problem with transporting masses of people, people who have low
> incomes. The solution we have come up with, the BRT, is worth other
> cities going emulate us. [...] I don't think any other country in this region, which applied this. And even to add: Successfully.

Arguably, the main objective of global BRT technocrats and consultants is to (re)produce the success of BRT and DART, which then materialises in the model's dissemination. ITDP's Africa programme director sees the possible replication of DART in the region as a form of success that would reflect back on and prove that DART itself was a success (ITDP 2018h). According to ITDP (2017c: 342–343), measuring and presenting success is very useful in order to ensure continuous learning and improvement, and to satisfy funding requirements. The success of various BRT systems has been translated into remarkable statistics. For example, Bogotá's *Transmilenio* boasts a 32 per cent reduction in travel time, a 93 per cent drop in bus accidents, and a 40 per cent drop in air pollution levels (Cervero 2005: 26; Hidalgo et al. 2013).

The desire to measure and declare the success of global BRT and its ma-terialisations contains two interrelated ambiguities. First, there is no uni-versal standard or clear definition of success or failure in terms of BRT; i.e. what is understood as success depends on subjective and normative as-sessments. For example, DART's service provider UDA-RT judges success in terms of revenue generation and reliable services. For BRT proponents, success is expressed in the daily ridership, the pphpd, the average travel time between terminals, the construction cost per kilometre, the overall length of the corridor or network and relatively low fares. Several categories work even better in selective comparison to other modes of transport, particularly minibus and light rail. The World Bank President argued that DART's two-thirds travel time reduction saved Dar es Salaam's commuters 16 days of travel every year – a 'great outcome for the people' (Kim in Lombard 2017). And according to a chairperson of the Prime Minister's Office, DART could daily save the country approximately 4.5 billion Tanzanian Shillings (equal to 2.4 million USD at that time) by reducing traffic congestion (DART Agency 2015). Representatives of the DART Agency and World Bank Tan-zania emphasised DART also benefitted private vehicle users by reducing the number of daladala and therefore congestion in mixed-traffic lanes. In short, a myriad of Dar es Salaam citizens benefit from DART because they have more time for work and leisure (GV 09/2016; JK 09/2016). Second, almost no BRT system has been declared unsuccessful let alone a failure. The self-assertion of a strong and coherent story helps to generate trust of the global BRT network and its members. The vast majority of politicians and planners do not look at how the global BRT model arrives at a specific site, and hence they miss its difficulties, controversies and tensions. This imbalance leads to an inherent contradiction of BRT. Since (the promise of) success is the key ingredient for global policy circulation, travelling models embody expectations, which are hard to be met.

Even before its inauguration and operation, DART has been marked for success. ITDP members, Tanzanian government officials and World Bank representatives frequently expressed high expectations throughout the pro-cess of what DART would achieve for the city, its citizens and its economy. ITDP prematurely announcement that DART was the first 'true' BRT and first *Gold Standard* in Africa (ITDP 2017d, 2018n; see also Chapter 3). Moreover, ITDP claims not only that DART will be 'truly transformative' by 'improving and transforming urban life in Tanzania' (ITDP 2016), but that it will also 'reverse congestion problems' and 'increase economic com-petitiveness', and that it will 'breathe new life' into the city and the East African region (ITDP 2018k: 19). The NGO also emphasises its own role in this project and presents DART as one of its main achievements in 2016. This demonstrates again the interdependency between ITDP and the global BRT model: if a BRT system offers high-quality service, ITDP has done high-quality consultancy work; conversely, if there was no demand for BRT, there would be no demand for ITDP's services either.

Both ITDP and the World Bank are keen to strengthen the position of global BRT in Africa, but expectations and pressure are also mounting due to the high amount of public and private debt that the project has produced. During a difficult period in 2014 (see Chapter 5), the then CEO of the DART Agency was apparently not concerned about failure: 'I never had any doubt that DART would be a success. [...] All implemented BRT systems are a success' (Tanzania Daily News 2014). In this context, the permanent secretary of the Prime Minister's Office emphasised: 'Making this a success is going to be of ultimate importance and will give crucial signals and lessons to future PPP endeavours. [...] Failure is not an option' (The Guardian Tanzania 2014). Due to DART's characteristic as a high-end model, another World Bank consultant stressed that 'DART has the chance to succeed [...], or let me put it differently: it has to succeed!' (BN 10/2015), and the first CEO of the DART Agency said: 'Nothing pays more than success. We are committed to work day and night to succeed despite the numerous challenges' (DART Agency 2014). 'The success of BRT cannot be downplayed', an ITDP employee told me while we were discussing DART's challenges, a particularly remarkable statement because it was the opposite of the impressions I had gathered in international settings of consultancy. No one, neither ITDP members nor any other powerful actor, openly sought to downplay DART's success; instead, everyone sought to downplay DART's difficulties. Equating 'DART' with 'success' meant that the success of DART became indisputable.

DART represents an opportunity to showcase the success of the BRT model; since 2016, the Tanzanian project has become probably the most-presented BRT system. DART appears on front covers of ITDP's publications, at local exhibitions in African cities, DART-specific webinars or in the shape of travelling DART stakeholders – and it is a prime destination for BRT study trips. Furthermore, ITDP presents the system as best practice regarding its institutional set-up, funding diversity and financial model (ITDP 2017c: 83, 2018h). However, due to DART's controversies and deviations from its initial plans, this global attention is also challenging. Because the global crowd must be convinced, stakeholders have tried to restrict controversies to BRT insiders, especially during early stages of DART's operations. After several months of operation, the then CEO of UDA-RT emphasised the pressure his company experienced: 'It's putting us under deep pressure to deliver what it was supposed to be. And that's a serious challenge' (HG 09/2016). Therefore, DART must meet various expectations that go beyond the mere improvement of public transport in Dar es Salaam. Like the famous Colombian BRT system, promises of positive side-effects of economic growth, environmental sustainability and social inclusion are inscribed into DART. In short, DART needs to succeed, and moreover, it needs to appear successful. As the model of African best practice, it bears the burden of fulfilling the imagination of Africa's urban future, characterised by economic sustainability, growth and stability.

As emphasised in the previous chapter, ITDP uses its tools like the STA and the *Mobilize* conference to promote itself. While emphasising its own role in the DART project, the NGO disguises its strong influence on determining the winners of the annual STA (ITDP 2018i, 2018j). Officially, an independent committee chooses the winning city, but ITDP has its ways to guide the decision making:

> I receive an email from a former ITDP colleague: 'I have a favor to ask, we are lobbying for Dar es Salaam to win the Sustainable Transport Award. Kindly nominate Dar Es Salaam Here. I have attached a sample filled-in nomination form. [...] The Sustainable Transport Award recognizes profound leadership and vision in sustainable transportation and urban livability, which we believe Dar Es Salaam has illustrated through the implementation of the BRT system.' I look at the attached form. All arguments listed, why Dar es Salaam should be the next winner of this award, refer to DART: 'The Dar es Salaam BRT is serving as a catalyst for the implementation of high-quality BRT systems in other African cities such as Nairobi, Kampala, Addis Ababa, and Kigali, all of which are in the process of implementing BRT systems.'
>
> (April 2017)

Two months later, out of ten nominated cities the committee awarded the STA to the Tanzanian metropolis (ITDP 2017a). With this strategic step, ITDP has come once again closer to the mutual success of itself and of DART, as well as the mutual constitution of BRT and the sustainable transport ideal. Also in the committee's statement, DART is the main reason for the city's reward: DART has led to 'transformative improvements' that might 'accelerate the pace of change' (ITDP 2018f). At the STA ceremony in Washington DC, a member of the STA committee said that 'Africa, following the example of Dar es Salaam, can wake up to the revolution of BRT', and a World Bank employee described how proud he was for Tanzania 'to take the initiative and become pioneer'. Another World Bank consultant described DART as a 'great example of the progress being made in the East Africa region', and that 'the success of this system will hopefully inspire replication' (ITDP 2018g). East African journalists describe this award as representation of 'ambitious and innovative efforts to shape the world's future, thanks to the presence of a functioning rapid transit system in the city' (Kolumbia 2017), how DART 'made history' (McSheffrey 2017), and how Nairobi might now learn from Dar es Salaam how to improve transport (Okoth 2017).

Dar es Salaam benefits from this award in several regards. First, the city has become even more present in the global network of urban and transport planning. Second, it has become a 'member of the family', as the then director of ITDP put it (ITDP 2018g). Third, DART's significance has further increased to not only being the first 'true' African BRT, but it has also brought the first STA to an African city. Since the award is followed by the *Mobilize* summit, ITDP and DART's stakeholders directly disseminated the transport model to other African city leaders who were attending. Members of ITDP presented this event as a 'rare opportunity for them to share best practices with each other' (ITDP 2018k). James Kombe, responsible for the DART project from the World Bank's side, said that the award is a 'good incentive […] to keep the pressure and to raise the bar, to get much better systems in Africa' (ITDP 2018g). Indeed, it seems that delegations from Lagos, Lusaka, Addis Ababa, Kampala and Nairobi were impressed by DART and saw the potential to learn from an African city (ITDP 2018h; Nachilongo 2018). The *Mobilize* summit programme included talks and discussions on sustainable transport topics as well as site visits, which almost exclusively concerned DART. In order to convincingly present a strong image of DART, ITDP made several short-term improvements to the physical infrastructure and bus scheduling just before the conference (PH 11/2017). Bringing this strategic tool to Dar es Salaam shows again how central DART is in ITDP's work and for Tanzanian politics.

Magufulify DART

Policies might easily be promoted as a success because the success is linked to the successful governing of the ruling party (Lovell 2017: 13). Political programmes of national development and economic growth are politics of technoscience, whereby power relations materialise in infrastructures (Barry 2015; Mitchell 2002) so that infrastructures like BRT, effecting extensive land-use change, can become subject to public protest. But they also provide the opportunity for technocrats to prove their public engagement:

> Because transportation reform visibly changes the face and functionality of the city, it often becomes a hot-button issue and a powerful negotiating device in the political landscape. Politicians can use transportation projects as tools for gaining sufficient popularity to be elected to office.
>
> (Bassett and Marpillero-Colomina 2013: 143)

DART, too, has been strategically stage-managed on several political occasions, often accompanied by international technocrats. Even the signings of contracts became ceremonies in Dar es Salaam, giving a first glimpse of what the new infrastructure might bring to the country. Especially at times of elections, currently implemented policies and projects become a central topic of political debates. Because elections count in infrastructure planning and infrastructure planning counts for elections, it is not hard to see why the ruling party Chama Cha Mapinduzi (CCM) was keen to inaugurate DART – one of the country's most famous infrastructure projects – before the general election in 2015.[10] Consequently, inaugurating infrastructure projects shortly before elections might be a strategy of the Tanzanian government to win votes. Latour (1996: 16, 141) has also shown how French politicians of the 1980s loved the idea of *Aramis*, a symbol of modernisation, just as Tanzanian politicians of the 2000s loved the idea to modernising urban transport through DART.

> The DART corridor, still empty of buses, is decorated with posters for the general election. The yellow-green of the CCM dominates the plastered concrete walls and piers. Ironically, the CCM as the ruling party had just put a ban on mounting advertisement and any political content within the BRT infrastructure. Beyond the plastered posters, the CCM also mounted huge banners on designated billboards. The banners show Magufuli, at the time minister of works and presidential candidate for the CCM. The pictures' captions say *'Chaguo La Watu'* ('Choice of the People', see Figure 4.2) and *'Hapa Kazi Tu!'* ('Hands On!'). One banner, which is located right in front of DART's bus depot and UDA-RT's headquarters in Jangwani, shows Magufuli on a construction site. Right next to him stands a CCM party member who works for the bus operator. 'What a nice coincidence', employees tell me at UDA-RT's office.
>
> (October 2015)

Figure 4.2 Magufuli banner next to the corridor of DART: 'Choice of the People' (author's photo 10/2015).

Hapa kazi tu! – John Magufuli's motto is a perfect fit for the DART project and for Tanzanian infrastructure politics in general, in which he has been decisively active. In recent years, the CCM has realised several major road infrastructure projects in Dar es Salaam, such as flyovers, elevated roads and bridges. Deriving from infrastructural vocabulary, Magufuli carries nicknames like Bulldozer, Iron Fist and *Jembe* (hoe) that reflect his promise to improve Tanzania's infrastructures. By the time he became President, a wave of 'Maguphoria' swept across the country (Brankamp 2015). His primary campaign pledge was to fight corruption, reduce governmental expenses and invest the money in public infrastructures instead. Tanzanians created *twitter* hash tags like #*magufulify* and #*whatwouldmagufulido* to express their enthusiasm for the promised winds of change (Kubanda 2015). Similar to blogs with names like '5 Things I Learned from Enrique Peña-losa' (Bloomingrock 2018), international journalists wrote articles about '10 Ways Tanzania's (Awesome) New President Shows How to Lead an African Country' (SAPeople Staff Writer 2015).

After Magufuli became President in October 2015, he used DART to showcase the fulfilment of his election promises, especially the fight against corruption. However, Magufuli's anti-corruption campaign opened up big political dimensions that did not lead to a frictionless and rapid inauguration of DART. On the contrary, judicial disputes between the new Tanzanian leadership and UDA-RT followed, and the inauguration of DART was rescheduled again (see Chapter 5). One of his first measures was to suspend the chief of the Tanzania Revenue Authority (TRA), responsible for administering tax money. The port of Dar es Salaam had been at the centre of scandals because large quantities of imported goods entered the country without being taxed, including the buses and equipment that the Tanzanian bus operator imported (see also Hönke and Cuesta-Fernandez 2018; Msikula 2015). The investigation and subsequent restructuring of this authority under the new Magufuli government meant that UDA-RT could not get an exemption to clear the imported buses and the devices for the intelligent transport system (ITS). The company pleaded with the government to release the buses so that it could begin operations, and promised to pay the import duties in instalments because it did not have enough capital to pay all the taxes at once (AH 10/2015; GV 05/2016). However, under the surveillance of Magufuli, the TRA adhered to its new policy, even if this meant another delay of DART's interim service. After months of negotiations, UDA-RT took out another loan in order to pay the import duties so that the company could finally start operations and generate revenue in May 2016. Further suspensions of government officials followed in 2015 and 2016 due to abuse of position and embezzlement of public money. Among them was the then CEO of the DART Agency, who was suspended from office two weeks before the final deadline for launching the operations of the interim service provider (ISP). The suspension was due to irregularities in the procurement of the ISP and the alleged concealment of information from the Tanzanian government (Barasa 2015; Chibwete 2016). However, several insiders told me that it was not the CEO's personal fault because everyone in this position would have had acted the same way, and that the crucial decisions about the ISP contract were taken at higher levels (HK 05/2016; JB 09/2016; RL 05/2016). The then CEO became the victim because someone had to be held responsible for the irregularities of the ISP.

Magufuli's fame persisted throughout the early days of DART operations. Tanzanians called his politics *mwendokasi* – efficient and high speed – corresponding to DART and the idea of an accelerating Dar es Salaam. After one year of the ISP operations, Magufuli requested more transparency from the operators and ordered a detailed report on the revenue collected. He claimed that the public had a right to know 'whether the project was viable and worth their tax money' (Mirondo 2017). The President's engagement had been internationally acknowledged and the *New African* magazine awarded him the title 'African of the Year 2016' (Kabendera 2016; Versi et al. 2017). The World Bank also stood behind Magufuli's government, as the bank's President Jim Yong Kim assured the Tanzanian public. During Kim's visit to Dar es Salaam for the ceremonial laying of the foundation stone for the flyover at Ubungo in March 2017, Kim spoke in high terms of the reduction of government spending and the scaling up of investments in infrastructure:

> I'm here because leaders all over Africa and leaders all over the world told me that you have to go to Tanzania to see what president Magufuli and his team are doing to try to stop corruption and build nation. [...] I want you to know that the World Bank Group is going to stand with you. We believe in you and we will never let anyone say that Tanzania cannot do it.
>
> (Kim 2017)

However, a growing number of people have started questioning Magufuli's politics. Ever more stories circulate about his authoritarian leadership and how it endangers democratic principles like freedom of speech and freedom of press (Mwakideu 2016; Paget 2017). The CCM politician is not the 'breath of fresh air' he promised to be (Allison 2017). Moreover, even if he wanted to, he could hardly dissolve the deeply intermingled structures of Tanzania's political economy, e.g. between the Simon Group, Shirika la Usafiri Dar es Salaam (UDA) and leading government officials.[11]

DART has changed the face of the political landscape in Dar es Salaam in many ways. On the one hand it has spawned new political constellations and redistributed agency and influence on the one hand. On the other hand, it has caused controversy regarding the identities and responsibilities of the actors involved. New actors have appeared and the role of the state in the transport sector has changed significantly. After the privatisation of UDA and liberalisation of the public transport sector in late 1970s and early 1980s, DART stands for the re-engagement of the Tanzanian state in the public transport sector. Not only did the Ministry of Works (by that time led by Magufuli) borrow heavily from the World Bank to construct roads, stations and terminals, the government is responsible for DART's maintenance, which amounts to an indirect subsidisation. The government is responsible for DART's maintenance, which amounts to an indirect subsidisation. Moreover, the Tanzanian government coordinates between different stakeholders, enforces DART-specific traffic rules, oversees operations and collects a fee from operators for using the physical infrastructure. The DART Agency, run directly by the President's Office, executes these tasks. The DART Agency has its challenges coordinating and moderating between the World Bank, national and local government, the bus operator and the users of the system, who all pursue different interests: money lenders want to enforce their conditions, politicians need to get votes, business companies need to act profit-oriented, and transport users want a comfortable, reliable and affordable system. The World Bank's leadership, convinced that DART can only be successfully implemented with international guidance, has engaged international short- and long-term transport consultants (BN 03/2015).

While employees of the DART Agency feel they are subjected to excessive government control, UDA-RT staff see the DART Agency as the executive power of the government (GV 11/2016; RL 10/2015). Moreover, the name 'DART' is not used consistently. People in Dar es Salaam variously use the term to refer to the DART Agency, the BRT operations, the operator UDA-RT or the overall system. To clarify this point, the operator UDA-RT put sheets of paper at the terminals: 'UDA-RT IS THE SERVICE PROVIDER OF BRT'.[12] Also ITDP's *BRT Standard* scoring pointed to the shortcoming of DART's branding (ITDP 2018c). Furthermore, decision making processes within the DART project and the institutional coordination are challenging due to various government entities being involved. Several ministries with subordinate offices and authorities take part in the management, construction and maintenance of DART's physical infrastructure, whereby the DART Agency – and possibly the service provider in the future – maintains DART's stations and terminals (MN 03/2015; PF 10/2015). Consultants described the lack of capacity – inefficiencies, underfinancing, insufficient qualification and expertise – as another difficulty of the project. They attempt to create a central transport entity, the Dar es Salaam Urban Transport Authority (DUTA), to coordinate the various (around 17) transport-related institutions (HM 03/2015; JK 03/2015). After UDA-RT breached the *ISP Agreement* and guaranteed itself its long-term participation in DART operations (see Chapter 5), the chief technical advisor of the DART Agency linked the fact that no one intervened early enough to the weakness of the DART Agency. Remarkably, whereas almost everyone spoke negatively about DART's institutional capacity, ITDP globally advocates the DART organisational setup as best practice (ITDP 2018b, 2018h).

This chapter has investigated the normative and technopolitical dimensions of the BRT model in terms of technocrats' strategic practices of dissemination and persuasion. The rule, activities and power of the global BRT community – labelled as BRT-cracy – involve an interdependency between themselves and the model. Global BRT technocrats apply strategies like one-sided storytelling and sugar-coating to make city officials of the Global South believe in BRT. The transport model promises to be more than transport because it (presumably) increases economic growth and ecological sustainability, and strengthens human values like democracy and equality. The notion of success is a key ingredient for the model's rapid dissemination in the Global South, and the storytelling is highly performative and effective. DART, addressed as the African best practice by ITDP, is now following in the footsteps of the hitherto global best practice *Transmilenio*. For ITDP, the strategy to make success a self-fulfilling prophecy has been effective: DART and BRT are successful precisely because they are presented as success. Different dimensions of BRT's success – as the best and only option for cities of the Global South and as a successful mode of public transport– are indispensable for its existence. Since no stories of bad practice BRT circulate, each BRT story seems to be another indicator of the model's success. The development of DART also illustrates how the model is integrated into political landscapes on various scales, and how politicians like Magufuli have used such a prestigious infrastructure project for their political benefit, despite its manifold controversies and setbacks. Hence, the global image of success also has ramifications within national politics: politicians have put effort into DART's appearance of success because this reflects well on themselves and the Tanzanian metropolis. Consequently, the success inscribed in DART has become a symbol of progress for the city, the national government, Africa and the global BRT community.

Notes

1 Peñalosa, Bogotá, and *Transmilenio* co-produce themselves. Pineda, referring to Latour's principle of symmetry (1987), substantiates the co-production of the city and its public transport system because they mutually formed themselves (Pineda 2010: 133–137). Peñalosa's career is built not only upon BRT, but also on his efforts to transform Bogotá, and 'on how to build sustainable cities that can not only survive but thrive in the future' (TED 2013).

2 ITDP and Peñalosa have collaborated with further global organisations, most importantly Volvo Research and Educational Foundations (VREF) and the World Resource Institute (WRI, including EMBARQ). These organisations are present in the US and Latin America rather than in Africa.

3 As a major funder in Tanzania, the World Bank is highly interested in financing (transport) infrastructure projects. At the time of this conflict, the World Bank

had other infrastructure projects lined up which needed to appear to be successful investments.

4 In January 2017, the World Bank approved the loan for Phase 3 and 4 before an international operator had joined the PPP framework of DART (see Chapters 5 and 6).

5 *Wanaangalia zaidi, kabla ya kutoa fedha, wanaangalia performance ya Phase 1. […] Kwa sababu hakuna mtu anayetaka kuweka fedha kwenye project ambazo is not doing well.*

6 Almost half of the World Bank's loan of approximately 300 million USD for Phase 1 went to a European construction company, which brought its own machines and highly skilled staff; Tanzanians did low-paid construction work (BN 03/2015; JK 03/2015; see also World Bank 2014).

7 *Tutapata wageni wengi zaidi. Wataopen up, kuona how it works.*

8 *Amazing. […] Like US. Km tuko Ulaya vile yaan!!! Kwetu pazuri!!!*

9 *Kwa kawaida watoto wanaozaliwa ndani ya ndege hupewa zawadi ya kusafiri bure maishani mwao kwa ndege za Shirika husika, ndio maana sasa UDART wanafikiria zawadi ya kumpa Sarah na binti yake huyo mpya!*

10 It had been common practice for ceremonial launches to be held shortly before elections: the launch of DART's Phase 1 construction works took place in September 2010, right before the presidential election. Phase 1 was initially projected to start operations on that date, but then had to be postponed to 2012. At this alternative inaugural event, then President Kikwete promised that within the next five years (i.e. the next legislative period), the problem of congestion would decrease significantly (Domasa 2012; Mfinanga and Madinda 2016: 161). Hönke and Cuesta-Fernandez observed a similar nexus of events concerning the construction of the port near Bagamoyo (2017: 13).

11 The former public transport company UDA disposes of supportive political relationships, built up since independence. International consultants told me rumours that coalitions of high-ranking politicians and the Simon Group – the major shareholder of UDA – control the public transport sector in Dar es Salaam. This collaboration effectively enabled UDA to create UDART and to become a permanent operator of DART (for details, see Chapter 5).

12 *UDART NDIYO MTOA HUDUMA WA MABASI YAENDAYO HARAKA.*

References

Allison, S. 2017. "What Would Magufuli Do". Accessed on Institute for Security Studies (https://issafrica.org/iss-today/whatwouldmagufulido-were-just-finding-out). Published 27.09.2017, retrieved 17.04.2020.

Ardila-Gómez, A. 2004. *Transit Planning in Curitiba and Bogotá. Roles of Interaction, Risk, and Change.* Submitted to the Massachusetts Institute of Technology for the Degree of Ph.D. in Urban and Transportation Planning, September 2004.

Barasa, C. 2015. "Magufuli Steps in to Address Loopholes in Procurement Act". Accessed at: *Tanzania Daily News* (http://dailynews.co.tz/index.php/features/45574-magufuli-steps-in-to-address-loopholes-in-procurement-act). Published 31.12.2015, retrieved 11.11.2018.

Barry, A. 2015. "Infrastructural Times". In: H. Appel, N. Anand and A. Gupta (eds.), *The Infrastructure Toolbox.* Fieldsights, September 24. Accessed at *Cultural Anthropology* (https://culanth.org/fieldsights/series/the-infrastructure-toolbox). Published 24.09.2015, retrieved 17.04.2020.

Bassett, T. and A. Marpillero-Colomina 2013. "Sustaining Mobility. Bus Rapid Transit and the Role of Local Politics in Bogotá". *Latin American Perspectives* 40:2, 135–145.

Behrends, A., S.-J. Park and R. Rottenburg 2014. "Travelling Models. Introducing an Analytical Concept to Globalisation Studies". In: ibid. (eds.), *Travelling Models in African Conflict Management. Translating Technologies of Social Ordering.* Leiden: Brill, 1–40.

Behrens, R. and S. Ferro 2016. "Barriers to Comprehensive Paratransit Replacement". In: R. Behrens, D. McCormick and D. Mfinanga (eds.), *Paratransit in African Cities. Operations, Regulation and Reform.* Oxon, New York: Routledge, 199–220.

Berney, R. 2017. *Learning from Bogotá. Pedagogical Urbanism ad the Reshaping of Public Space.* Austin: University of Texas Press.

Björkman, L. and A. Harris 2018. "Engineering Cities: Mediating Materialities, Infrastructural Imaginaries and Shifting Regimes of Urban Expertise". *International Journal of Urban and Regional Research* 42:2, 244–262.

Bloomingrock 2018. "5 Things I Learned from Enrique Peñalosa". Accessed at *Smart Cities Dive* (http://www.smartcitiesdive.com/ex/sustainablecitiescollective/5-things-i-learned-enrique-pe-alosa/1010986/), retrieved 17.04.2020.

Boje, D. 1991. "The Storytelling Organization. A Study of Story Performance in an Office-Supply Firm". *Administrative Science Quarterly* 36:1, 106–126.

Bowker, G. 2015. "Temporality". In: H. Appel, N. Anand and A. Gupta (eds.), *The Infrastructure Toolbox.* Fieldsights, September 24. Accessed at *Cultural Anthropology* (https://culanth.org/fieldsights/series/the-infrastructure-toolbox). Published 24.09.2015, retrieved 17.04.2020.

Boyer, D. 2008. "Thinking Through the Anthropology of Experts". *Anthropology in Action* 15:2, 38–46.

Brankamp, H. 2015. "#WhatWouldMagufuliDo Sparks New Bout of Tanzania". Accessed at *African Arguments* (http://africanarguments.org/2015/11/30/-whatwouldmagufulido-sparks-new-bout-of-tanzaphilia/). Published 30.11.2015, retrieved 17.04.2020.

Callon, M., P. Lascoumes and Y. Barthe 2009 (2001). *Acting in an Uncertain World. An Essay on Technical Democracy.* Cambridge, MA: MIT Press.

Cameron, E. 2012. "New Geographies of Story and Storytelling". *Progress in Human Geography* 36:5, 573–592.

Cervero, R. 2013. "Bus Rapid Transit (BRT). An Efficient and Competitive Mode of Public Transport". Working Paper 2013-01, University of California, Institute of Urban and Regional Development.

Cervero, R. 2005. "Progressive Transport and the Poor. Bogotá's Bold Steps Forward". *Access* 27, 24–30.

Chibwete, R. 2016. "Simbachawene Appoints New Dart Executive". Accessed at *The Citizen Tanzania* (http://www.thecitizen.co.tz/News/Simbachawene-appoints-new-Dart-executive/-/1840340/3022730/-/3ryqy9z/-/index.html). Published 06.01.2016, retrieved 17.04.2020.

Clarence, E. 2002. "Technocracy Reinvented. The New Evidence Based Policy Movement". *Public Policy and Administration* 17:3, 1–11.

Czarniawska, B. 2004. *Narratives in Social Science Research.* London, Thousand Oaks, CA: Sage.

DART Agency 2015. "Parliament Committee Want BRT Project Kick Off". Accessed at *DART Agency* (http://dart.go.tz/en/parliamentary-committee-want-brt-project-kick-off/). Published March 2015, retrieved 20.07.2016.

DART Agency 2014. "Welcoming Statement: CEO of DART". Accessed at *DART Agency* (http://dart.go.tz/en/chief-executive-officers-welcoming-statement/). Published June 2014, retrieved 20.07.2016.

De Laet, M. and A. Mol 2000. "The Zimbabwe Bush Pump. Mechanics of a Fluid Technology". *Social Studies of Science* 30:2, 225–263.

Domasa, S. 2012. "Project to Decongest Dar Launched". Accessed at *The Guardian Tanzania* (http://www.ippmedia.com/frontend/index.php?l=46008). Published 20.09.2012, retrieved 11.11.2018.

Edwards, P. and G. Hecht 2010. "History and the Technopolitics of Identity. The Case of Apartheid South Africa". *Journal of Southern African Studies* 36:3, 619–639.

Flores, O. 2011. "La Receta Transmilenio". Accessed at *AnimalPolitico* (https://www.animalpolitico.com/blogueros-ciudad-posible/2011/01/29/la-receta-transmilenio/). Published 29.01.2011, retrieved 17.04.2020.

Graham, S. and N. Thrift. 2007. "Out of Order. Understanding Repair and Maintenance". *Theory, Culture & Society* 24:3, 1–25.

Grijzen, J. 2010. *Outsourcing Planning. What Do Consultants Do in Regional Spatial Planning in the Netherlands*. Amsterdam: University of Amsterdam Press.

Harvey, P. and H. Knox 2015. *Roads. An Anthropology of Infrastructure and Expertise*. Ithaca, London: Cornell University Press.

Healey, P. 2012. "The Universal and the Contingent. Some Reflections on the Transnational Flow of Planning Ideas and Practices". *Planning Theory* 11:2, 188–207.

Hecht, G. 2011. "Introduction". In: ibid. (ed.), *Entangled Geographies. Empire and Technopolitics in the Global Cold War*. Cambridge, MA: MIT Press, 1–12.

Hensher, D. 2007. "Sustainable Transport Systems. Moving Towards a Value for Money and Network-Based Approach and Away from Blind Commitment". *Transport Policy* 14, 98–102.

Hidalgo, D., L. Pereira, N. Estupiñán and P. Jiménez 2013. "TransMilenio BRT System in Bogotá, High Performance and Positive Impact. Main Results of an Ex-Post Evaluation". *Research in Transportation Economics* 39, 133–138.

Hönke, J. and I. Cuesta-Fernandez 2017. "A Topolographical Approach to Infrastructure. Political Geography, Topology and the Port of Dar es Salaam". *Environment and Planning D*, 1–20.

Höhnke, C. 2012a. "Potentials and Pitfalls of Transport Innovations. Lessons from Santiago de Chile". *Trialog* 110, 13–16.

Höhnke, C. 2012b. *Verkehrsgovernance in Megastädten. Die ÖPNV-Reformen in Santiago de Chile und Bogotá*. Stuttgart: Franz Steiner Verlag.

Hönke, J. and I. Cuesta-Fernandez 2017. "A Topolographical Approach to Infrastructure. Political Geography, Topology and the Port of Dar es Salaam." *Environment and Planning D*, 1–20.

Hook, W. 2009. "Bus Rapid Transit. A Cost-Effective Mass Transit Technology". *EM* 26–32.

Howe, C., J. Lockrem, H. Appel, E. Hackett, D. Boyer, R. Hall, M. Schneider-Mayerson, A. Pope, A. Gupta, E. Rodwell, A. Ballestero, T. Durbin, F. el-Dahdah, E. Long and C. Mody 2015. "Paradoxical Infrastructures. Ruins, Retrofit, and Risk". *Science, Technology, & Human Values* 41:3, 1–19.

ITDP 2020. "History of ITDP". Accessed at *ITDP* (https://www.itdp.org/who-we-are/history-of-itdp/). Retrieved 17.04.2020.

ITDP 2018a. "About ITDP". Accessed at *ITDP* (https://www.itdp.org). Retrieved 11.11.2018.

ITDP 2018b. "BRT Planning 101". Accessed at *ITDP* (https://www.itdp.org/publication/webinar-brt-planning-101/). Published 02.02.2018, retrieved 17.04.2020.

ITDP 2018c. "BRT Planning 301: Communications". Accessed at *ITDP* (https://www.itdp.org/publication/webinar-brt-planning-301/). Published 03.04.2018, retrieved 17.04.2020.

ITDP 2018d. "BRT Planning 501: Integration". Accessed at *ITDP* https://www.itdp.org/2018/10/09/webinar-brt-501-integration/). Published 09.10.2018, retrieved 17.04.2020.

ITDP 2018e. "Dar es Salaam Leads a Breakthrough for African Cities". *Sustainable Transport*, Winter 2018, No. 29. Accessed at *ITDP* (https://www.itdp.org/category/content-type/publication/magazine/). Published 13.02.2018, retrieved 11.11.2018.

ITDP 2018f. "2018: Dar es Salaam, Tanzania". Accessed at *STA* (http://staward.org/winners/2018-dar-es-salaam-tanzania/). Retrieved 17.04.2020.

ITDP 2018g. "Dar es Salaam, Tanzania Presented with 2018 Sustainable Transport Award". Accessed at *ITDP* (https://www.itdp.org/2018-sta-ceremony/). Published 30.01.2018, retrieved 17.04.2020.

ITDP 2018h. "DART: Transforming Urban Mobility in East Africa". Accessed at *ITDP* (https://www.itdp.org/2018/06/01/webinar-dart-transforming-mobility/). Published 01.06.2018, retrieved 17.04.2020.

ITDP 2018i. "In Dar es Salaam, African Leader Reflect on MOBILIZE". Accessed at *ITDP* (https://www.itdp.org/2018/08/07/african-leaders-reflect-mobilize/). Published 07.08.2018, retrieved 17.04.2020.

ITDP 2018j. "MOBILIZE". Accessed at *Mobilize Summit* (https://mobilizesummit.org). Retrieved 11.11.2018.

ITDP 2018k. "Sustainable Transport Bulletin. Summer 2018". Accessed at *MailChimp* (https://mailchi.mp/itdp/44x03kie08-1006355?e=2864d0af9a). Published 02.08.2018, retrieved 17.04.2020.

ITDP 2018l. "What Is BRT"? Accessed at *vimeo* (https://vimeo.com/276223930). Published 21.06.2018, retrieved 17.04.2020.

ITDP 2018m. "What We Do: Public Transport". Accessed at *ITDP* (https://www.itdp.org/what-we-do/public-transport/). Retrieved 17.04.2020.

ITDP 2018n. "Where We Work: Africa". Accessed at *ITDP* (https://www.itdp.org/-where-we-work/Africa). Retrieved 11.11.2018.

ITDP 2017a. "Dar es Salaam, Tanzania Wins 2018 Sustainable Transport Award". Accessed at *ITDP* (https://www.itdp.org/dar-es-salaam-wins-2018-sta/). Published 02.07.2017, retrieved 17.04.2020.

ITDP 2017b. "Santiago, Chile Putting Pedestrians First". *Sustainable Transport*, Winter 2017, No. 28. Accessed at *ITDP* (https://www.itdp.org/category/content-type/publication/magazine/). Published 02.02.2017, retrieved 11.11.2018.

ITDP 2017c. "The Online BRT Planning Guide". 4th Edition. Accessed at *ITDP* (https://brtguide.itdp.org). Published 16.11.2017, retrieved 17.04.2020.

ITDP 2017d. "What Can Cities Do to Ensure a Better Climate and Better Future for Everyone"? Accessed at *MailChimp* (https://mailchi.mp/itdp/what-can-cities-do-to-ensure-a-better-future-for-everyone?e=2864d0af9a). Published 12.12.2017, retrieved 17.04.2020.

ITDP 2016. "How DART Could Transform Urban Life in Tanzania". Accessed at *Campaign Archive* (https://us7.campaign-archive.com/?u=0b5e10eda9e3afdb7 eceb76f6&id=8993bb718d&e=2864d0af9a09.08). Published 08.09.2016, retrieved 17.04.2020.

ITDP Africa 2016. "ITDP Africa". Accessed at *twitter* (https://twitter.com/itdpafrica?lang=de). Published 22.11.2016, retrieved 17.04.2020.

ITDP 2015a. "30 Years of ITDP". Accessed at *ITDP* (www.itdp.org/30-years-of-itdp). Published 01.01.2015, retrieved 17.04.2020.

ITDP 2015b. "ITDP Announces the Resignation of CEO Walter Hook after 21 Years". Accessed at *ITDP* (https://www.itdp.org/itdp-announces-resignation-ceo-walter-hook-21-years/). Published 05.01.2015, retrieved 17.04.2020.

ITDP 2003. "Enrique Peñalosa Inspires Change Across Africa". Accessed at *ITDP* (https://www.itdp.org/enrique-penalosa-inspires-change-across-africa/). Published 01.02.2002, retrieved 17.04.2020.

Jaffe, E. 2015. The Myth That Everyone Naturally Prefers Trains to Buses". Accessed at *CityLab* (https://www.citylab.com/transportation/2015/02/the-myth-that-everyone-naturally-prefers-trains-to-buses/385759/?utm_source=SFTwitter). Published 23.02.2015, retrieved 17.04.2020.

Kabendera, E. 2016. "Tanzania's President Magufuli. Man of the People, Man of the Party"? Accessed at *African Arguments* (http://africanarguments.org/2016/08/16/tanzanias-president-magufuli-man-of-the-people-man-of-the-party/). Published 16.08.2016, retrieved 17.04.2020.

Kim, J. 2017. "Laying Foundation Stone for Ubungo Overpass Speech. Remarks by Dr Jim Yong Kim, President of the World Bank Group". Accessed at *Ikulu* (http://blog.ikulu.go.tz/?p=20114). Published 20.03.2017, retrieved 17.04.2020.

Kolumbia, L. 2017. "Dar Scoops World City Mobility Award". Accessed at *The Citizen Tanzania* (http://www.thecitizen.co.tz/News/Dar-scoops-world-city-mobility-award/1840340-4216836-128wj00/index.html). Published 06.12.2017, retrieved 17.04.2020.

Kubanda, F. 2015. "Magufulify". Accessed at *twitter* (https://twitter.com/hashtag/magufulify). Published 24.11.2015, retrieved 11.11.2018.

Lagendijk, A. and S. Boertjes 2012. "Light Rail: All Change Please! A Post-Structural Perspective on the Global Mushrooming of a Transport Concept". *Planning Theory* 12:3, 290–310.

Larner, W. and N. Laurie 2010. "Travelling Technocrats, Embodied Knowledges. Globalising Privatisation in Telecoms and Water". *Geoforum* 41, 218–226.

Latour, B. 1996. *Aramis or the Love of Technology*. Cambridge, MA: Harvard University Press.

Latour, B. 1987. *Science in Action. How to Follow Scientists and Engineers through Society*. Cambridge, MA: Harvard University Press.

Latour, B. 1983. "Give Me a Laboratory and I Will Raise the World". In K. Knorr-Cetina and M. Mulkay (eds.), *Science Observed. Perspectives on the Social Study of Science*. Los Angeles, CA: Sage, 141–170.

Lombard, H. 2017. "Time is Money. Transforming Dar es Salaam's Road Transport to Reduce Dense Traffic". Accessed on *World Bank* (http://www.worldbank.org/en/news/feature/2017/03/27/time-is-money-transforming-dar-es-salaams-road-transport-to-reduce-dense-traffic?cid=ECR_E_NewsletterWeekly_EN_EXT). Published 27.03.2017, retrieved 17.04.2020.

Lovell, H. 2017. "Policy Failure Mobilities". *Progress in Human Geography* 43:1, 1–18.

Mawdsley, E. 2017. "Development Geography I. Cooperation, Competition and Convergence between 'North' and 'South'". *Progress in Human Geography* 41:1, 108–117.

McCann, E. and K. Ward 2013. "A Multi-Disciplinary Approach to Policy Transfer Research. Geographies, Assemblages, Mobilities and Mutations". *Policy Studies* 34:1, 2–18.

McCann, E. 2011. "Urban Policy Mobilities and Global Circuits of Knowledge. Toward a Research Agenda". *Annals of the Association of American Geographers* 101:1, 107–130.

McSheffrey, E. 2017. "Tanzania's Dar es Salaam Makes History for Sustainable Transit Award". Accessed at *National Observer Canada* (http://www.national observer.com/2017/07/13/news/tanzanias-dar-es-salaam-makes-history-sustainable-transit-award). Published 13.07.2017, retrieved 17.04.2020.

Merry, S. 2006. "Translational Human Rights and Local Activism. Mapping the Middle". *American Anthropologist* 108:1, 38–51.

Meynaud, J. 1964. *Technocracy*. London: Faber and Faber.

Mfinanga, D. and E. Madinda 2016. "Public Transport and Daladala Service Improvement Prospects in Dar es Salaam". In: R. Behrens, D. McCormick and D. Mfinanga (eds.), *Paratransit in African Cities. Operations, Regulation and Reform.* Oxon, New York: Routledge, 155–173.

Mirailles, J.-M. 2012. "Mass Transit Modes Relevance in Developing Countries. The Case of Bogotá". *Trialog* 110, 8–12.

Mirondo, R. 2017. "Simon Group Pays for Lion's Share in UDA". Accessed at *The Citizen Tanzania* (http://www.thecitizen.co.tz/News/Simon-Group-pays-for-lions-share-in-UDA/1840340-3787698-taxxfc/index.html). Published 26.01.2017, retrieved 17.04.2020.

Mitchell, T. 2002. *Rule of Experts. Egypt, Techno-Politics, Modernity.* Berkeley: University of California Press.

Msigwa, R. 2013. "Challenges Facing Urban Transportation in Tanzania". Mathematical Theory and Modelling 3:5, 18–26.

Msikula, A. 2015. "Why UDA-RT Bus Project is Stuck. Sh8 Billion Import Duty Stalls Kick-Off". Accessed at *The Guardian Tanzania* (http://www.ippmedia.com/?l=87195). Published 20.12.2015, retrieved 11.11.2018.

Mwakideu, C. 2016. "Magufuli's 'Bulldozing' Leadership Questioned". Accessed at *Deutsche Welle* (http://www.dw.com/en/magufulis-bulldozing-leadership-questioned/a-19371737). Published 01.07.2016, retrieved 17.04.2020.

Nachilongo, H. 2018. "Lagos Delegation Applauds Tanzania for Rapid Transport Services". Accessed at *The Citizen Tanzania* (http://www.thecitizen.co.tz/News/Lagos-delegation-applauds-Tanzania-for-rapid-transport-services/1840340-4642788-11h48f0z/index.html). Published 02.07.2018, retrieved 17.04.2020.

Okoth, E. 2017. "Lessons from Dar on Unclogging City Traffic Jams". Accessed at *Business Daily Africa* (http://www.businessdailyafrica.com/corporate/Lessons-from-Dar--on-unclogging-city-traffic-jams-/539550-3979980-g2v0ml/index.html?platform=hootsuite). Published 20.06.2017, retrieved 17.04.2020.

Paget, D. 2017. "Tanzania's Anti-Corruption Crusader Cracks Down on Opponents". Accessed on *CNN* (https://edition.cnn.com/2017/11/07/africa/magufuli-crackdown/index.html). Published 07.11.2017, retrieved 17.04.2020.

Peck, J. and N. Theodore 2015. *Fast Policy. Experimental Statecraft at the Thresholds of Neoliberalism*. Minneapolis, London: University of Minnesota Press.

Peck, J. and N. Theodore 2010. "Mobilizing Policy. Models, Methods, and Mutations". *Geoforum* 41, 169–174.

Peñalosa, E. 2013. "Why Buses Represent Democracy in Action". Accessed at *TED* (https://www.ted.com/talks/enrique_penalosa_why_buses_represent_democracy_in_action#t-221029). Published September 2013, retrieved 17.04.2020.

Peñalosa n.d. "A Developed Country". Accessed at *TransitApp* (http://shitenriquesays.com/a-developed-country), retrieved 17.04.2020.

Pineda, A. 2010. "How Do We Co-Produce Urban Transport Systems and the City? The Case of Transmilenio and Bogotá". In: I. Farías and T. Bender (eds.), *Urban Assemblages. How Actor-Network Theory Changes Urban Studies*. Oxon, New York: Routledge, 123–138.

Prince, R. 2016. "The Spaces in between. Mobile Policy and the Topographies and Topologies of the Technocracy". *Environment and Planning D* 34:3, 420–437.

Prince, R. 2012. "Policy Transfer, Consultants and the Geographies of Governance". *Progress in Human Geography* 36:2, 188–203.

Project for Public Spaces 2008. "Enrique Peñalosa". Accessed at *PPS* (https://www.pps.org/reference/epenalosa-2/). Published 31.12.2008, retrieved 17.04.2020.

Rizzo, M. and A. Dave 2018. "Questioning BRTs: A Win-Win Solution to Public Transport Problems in the Cities of Developing Countries"? Accessed at *SOAS* (https://www.soas.ac.uk/registry/scholarships/file126686.pdf). Published 13.02.2018, retrieved 17.04.2020.

Rizzo, M. 2017. *Taken for a Ride. Grounding Neoliberalism, Precarious Labour and Public Transport in an African Metropolis*. Oxford: Oxford University Press.

Roy, A. 2012. "Ethnographic Circulations. Space – Time Relations in the Worlds of Poverty Management". *Environment and Planning A* 44:1, 31–41.

Roy, A. 2005. "Urban Informality. Toward an Epistemology of Planning". *Journal of the American Planning Association* 71:2, 147–158.

SAPeople Staff Writer 2015. "10 Ways Tanzania's (Awesome) New President Shows How to Lead an African Country". Accessed at *SAPeopleNews* (http://www.sa-people.com/2015/11/29/10-ways-tanzanias-awesome-new-president-shows-how-to-lead-an-african-country/). Published 29.11.2015, retrieved 17.04.2020.

Schwanen, T. 2016. "Geographies of Transport I. Reinventing a Field"? *Progress in Human Geography* 40:1, 126–137.

Schwelger, T. and M. Powell 2008. "Unruly Experts. Methods and Forms of Collaboration in the Anthropology of Public Policy". *Anthropology in Action* 15:2, 1–9.

Shapin, S. 2008. "Science and the Modern World". In: E. Hackett, O. Amsterdamska, M. Lynch and J. Wajcman (eds.), *The Handbook of Science and Technology Studies (Third Edition)*. Cambridge, MA: MIT Press, 433–448.

Stein, C., B. Michel, G. Glasze and R. Pütz 2017. "Learning from *Failed* Policy Mobilities: Contradictions, Resistances and Unintended Outcomes in the Transfer of 'Business Improvement Districts' to Germany". *European Urban and Regional Studies* 24:1, 35–49.

Stirrat, R. 2000. "Cultures of Consultancy". *Critique of Anthropology* 20:1, 31–46.

Stone, D. 2004. "Transfer Agents and Global Networks in the 'Transnationalization' of Policy". *Journal of European Public Policy* 11:3, 545–566.

Tanzania Daily News 2014. "Dart Project Will Be a Success – Mlambo". Accessed at *Tanzania Daily News* (https://allafrica.com/stories/201401270478.html). Published 26.01.2014, retrieved 17.04.2020.

TED 2013. "Enrique Peñalosa. Colombian Politician, Urban Activist". Accessed at *TED Global* (https://www.ted.com/speakers/enrique_penalosa). Published December 2013, retrieved 17.04.2020.

Temenos, C. and E. McCann 2013. "Geographies of Policy Mobilities". *Geography Compass* 7:5, 344–357.

Temenos, C. and E. McCann 2012. "The Local Politics of Policy Mobility. Learning, Persuasion, and the Production of a Municipal Sustainability Fix". *Environment and Planning A* 44:6, 1389–1406.

The Guardian Tanzania 2014. "Dar Life to Brighten Up when DART Becomes Fully Operational". Accessed at *The Guardian Tanzania* (http://www.ippmedia.com/frontend/index.php?l=63983). Published 22.01.2014, retrieved 11.11.2018.

UDA 2017. *UDA Leo*. Toleo na. 07. July 2017.

Ureta, S. 2015. *Assembling Policy. Transantiago, Human Devices, and the Dream of a World-Class Society*. Cambridge, MA: MIT Press.

Versi, A., E. Kabendera and N. Ford 2017. "John Magufuli: Tanzania's Rising Star". Accessed at *New African Magazine* (http://newafricanmagazine.com/john-magufuli-tanzanias-rising-star/). Published 08.06.2017, retrieved 17.04.2020.

Ward, K. 2011. "Policies in Motion and in Place. The Case of Business Improvement Districts". In: E. McCann and K. Ward (eds.), *Mobile Urbanism. City and Policymaking in the Global Age*. Minneapolis: University of Minnesota Press, 71–95.

Wilson, G. 2006. "Beyond the Technocrat? The Professional Expert in Development Practice". *Development and Change* 37:3, 501–523.

Wood, A. 2016. "Tracing Policy Movements. Methods for Studying Learning and Policy Circulation". *Environment and Planning A* 48:2, 391–406.

Wood, A. 2015a. "Multiple Temporalities of Policy Circulation. Gradual, Repetitive and Delayed Processes of BRT Adoption in South African Cities". *International Journal of Urban and Regional Research* 39:3, 1–13.

Wood, A. 2015b. "The Politics of Policy Circulation. Unpacking the Relationship between South African and South American Cities in the Adoption of Bus Rapid Transit". *Antipode* 47:4, 1062–1079.

Wood, A. 2014. "Moving Policy. Global and Local Characters Circulating Bus Rapid Transit through South African Cities". *Urban Geography* 35:8, 1238–1254.

World Bank 2017. "Projects and Operations. Project. Dar es Salaam Urban Transport Improvement Project". Accessed at *The World Bank* (http://projects.worldbank.org/P150937?lang=en). Published 08.03.2017, retrieved 17.04.2020.

World Bank 2014. *International Development Association. The World Bank's Fund for the Poorest*. Washington, DC: Development Finance Vice Presidency of the World Bank.

Wright, L. and W. Hook 2007. *Bus Rapid Transit Planning Guide*. New York: ITDP.

5 Effective operational controversies

Two weeks after UDA-RT has started interim services, I see UDA-RT staff affixing stickers on DART's buses. Coming closer, I realise it is the logo of the DART Agency, scarcely legible cyan letters on the sky-blue vehicle body. The logo stands in contrast to the one of UDA-RT, shining in red and white from the buses' shell. Whereas a manager of the bus manufacturer explains to me that the company was only tasked with the UDA-RT logo, the official *Bus Output Specifications* document states that UDA-RT had to cater for the visibility of DART on the buses (DART Agency 2014a: 13). UDA-RT management tried to circumvent this specification and did not forward it to the manufacturer. Only now, after the DART Agency told the company to do so, is UDA-RT affixing the DART logo. Without the DART logo, UDA-RT would not have been allowed to use the buses on the government's corridor. Only a week later, I see the provisionally affixed DART logos flaking off, whereas the UDA-RT letters keep on shining on the vehicle bodies.

(May 2016)

Seeking to understand the (re)assembling of a global model and how actors deal with its adaptation (see Akrich 1992; Latour 1987), this chapter is concerned with controversies within DART's operational model and the strategies of the bus operators, Usafiri Dar es Salaam Rapid Transit (UDA-RT), to change the model. The World Bank's requirement to have a competitive international tender for DART operations led to great controversies in Dar es Salaam. UDA-RT, under the lead of its major shareholder UDA, tried to circumvent the requirements by not complying with them. The affixing of stickers illustrates in a nutshell the underlying controversy between the World Bank, the DART Agency and UDA-RT, who have different understandings of responsibility and reliability, and competing interests of participation and ownership. Transport actors in Dar es Salaam responded to the global BRT model by rejecting the imposed operational design, which is predominantly stipulated by global consultants. Adopting various strategies with the indirect support of political elites, transport operators succeeded in reshaping DART. They reformulated power relations in Dar es Salaam by claiming a permanent role in the model of DART's operational structure.

Telling the story of how the 'DART model' became the 'UDA-RT model' and how an interim service provider succeeded to receive a permanent status reveals the connectedness of Tanzania's transport sector and government. Since power is unequally distributed in assemblages (Farías 2011; Ureta 2014), this chapter demonstrates that infrastructure is political in terms of its 'possibility of conflict and disagreement' (Barry 2015). Moreover, it demonstrates that contradiction and contestation are inherent to gradual planning and implementation processes (Moore 2013; Wood 2015b), so that the circulation and translation of policies are characterised by friction, contingency and conflict, as well as uncertainty. Shifting the focus from the ways that global consultants, models and places elsewhere shape the local, I now scrutinise how policies are assembled and 'made up locally' (Robinson 2015: 831; Temenos and Baker 2015).

During my research, various DART stakeholders assured me that planning and building a BRT system was comparably easy. The former chief technical advisor of the DART Agency pointed out: 'Building is simple, anybody can build it. But managing, operating it [is a] power game' (HM 03/2015). Similarly, the head of the construction company's technical office emphasised: 'The construction is not particularly exciting; it's not a nuclear power plant. We just build it. [...] The exciting part is its implementation'[1] (MH 09/2015). The engineers were right, even if planning and constructing DART was not completely free of controversy (see Rizzo 2017). The director of operations and infrastructure management of the DART Agency underlined that, in order to become a world-class BRT, it was important for DART's operations to attain the same high quality as its physical infrastructure (RL 10/2015). However, after the BRT model had been put into socio-material practice, conflicts between stakeholders regarding the system's operations have led to various unplanned modifications of the designs. This illustrates that resistance to BRT is more likely to occur when the initiative has not evolved out of the local context (Behrens and Ferro 2016: 201–202; Vigar 2017: 41). Apparently aware of this issue, ITDP (2018a) quotes a Chinese proverb in a webinar: 'To open a shop is easy, but to keep it open is an art'. The NGO warns particularly against possible controversies with local operators:

> The most difficult negotiations for developing a BRT system will likely be with existing transit operators. Change is never easy and, regardless of the benefits, many will resist even if they are brought on board early and well.

> (ITDP 2017: 330)

I refer to controversies as 'disputes that emerge around technical matters, especially those that question technical fixes and standard narratives about infrastructure' (Hönke and Cuesta-Fernandez 2018: 247). Controversies have porous boundaries due to the varying fields of expertise involved (Pinch and Leuenberger 2006: 6), meaning that the (technical) system has not stabilised at the time controversies evolve. As Callon et al. (2009) have argued, a socio-technical controversy can act as an effective and powerful apparatus for actors to explore social and technical uncertainties in hybrid forums. Resolving controversies in such forums can possibly have a democratising influence because controversies strengthen public participation and activate citizens. Controversies may thus enrich democracy and clarify the meaning of a conflictual situation, making them powerful apparatuses for exploring and learning about possible worlds. Therefore, controversies are effective in order to develop new forums, include other actors and contribute to a dialogic democracy (ibid.: 30, 99). This means that controversies, appearing in moments of uncertainty and disagreement, and debated between experts and laypersons, do not necessarily hinder political planning processes. Rather, they

> encourage the enrichment and transformation of the initial projects and stakes, simultaneously permitting the reformulation of problems, the discussion of technical options, and, more broadly, the redefinition of the objectives pursued.
>
> (ibid. 2009: 32)

Accordingly, Collier and Ong argue that the ways in which actors call global assemblages into question provides insight into the nature of global forms:

> They [global assemblages] are domains in which the forms and values of individual and collective existence are problematized or at stake, in the sense that they are subject to technological, political, and ethical reflection and intervention.
>
> (2005: 4)

Hence, examining controversies provides a fruitful perspective on political decision making and its uncertainties, the temporality of infrastructures, and on economic and social dimensions of infrastructural processes and the technological in general (Barry 2012, 2013). However, researchers need to go 'beyond the usual naïve claims about the benefits of "diversity" and "participation"' (Ureta 2015: 168). Appreciating democracy – in the sense of participation – is not sufficient because participation also means exercising power. Instead of criticising technocracy and discussing 'the real needs versus utopian schemes' (ibid.: 162–163), researchers need to acknowledge the complexity and heterogeneity of (non)human actors within such a planning endeavour that always involves struggle and requires coordination within these assemblages.

Controversies and power relations are intertwined; they necessarily raise the question of who decides and speaks for whom. Controversies are also not necessarily highly visible; they do not always involve the public or confront the (attempted) closure of knowledge. Located between stabilisation and destabilisation of inherited norms and knowledges, controversies can be complex and multiple, and several controversies can overlap, complement or contradict each other. To date, the literature has mainly focussed on planning controversies and the mobilisation of best practices, and has insufficiently taken controversies regarding the implementation of policies and operating infrastructures into account. In STS scholarship, a number of approaches to socio-technical (knowledge) controversies have evolved since scientists became increasingly involved in political areas (Nelkin 1975, 1979). In the last decade, studies of controversies have opened up to new methodologies, diverse forms of controversies (e.g. technopolitical controversies), as well as to non-experts and lay knowledge (c.f. Marres 2015; Venturini 2009; Whatmore 2009). The articulation of technoscientific issues by the public, which Marres (2007, 2012) calls 'issue formation', gives rise to knowledge controversies. Hence, the settlement of a controversy leads to a society's stability and not vice versa (Latour 1987; see also Jasanoff 2012; Stengers 2005). In this regard, I explore the unpredictability of technopolitical processes and the power relations that BRT controversies bring out.

BRT controversies

The most prominent stories of protests against BRT derive from Delhi and Johannesburg. Delhi's BRT has the dubious distinction of being the only BRT worldwide to have shut down after inauguration. The reason: the city's (mainly middle-class) citizens claimed the BRT corridor for mixed-traffic because they could see no benefit from this new transport system (AM 09/2015; see also Hidalgo 2015; Msira 2016). In Johannesburg, during the first weeks of BRT operation, each bus had to be escorted by police because minibus drivers were shooting at them. Further stories of controversial BRT systems include Santiago de Chile, where the national government aimed to restore the credibility of *Transantiago* after the emergence of public controversies regarding ticket prices and poor implementation of the system (Ureta 2015), *Transmilenio*'s crowded buses and high fares (Jaffe 2012), and the uncertainty of ownership and operational arrangements in Nelson Mandela Bay (Wood 2015a). Despite this, stories of resistance, controversy and failure remain largely absent in the global BRT narrative.

Even if the underlying controversies were similar to disputes regarding participation and power elsewhere, the resistance to DART has been different, and more strategic, than in other cities. Neither the Tanzanian population nor daladala operators engaged in direct boycotts when BRT operations started in 2016. Around 2010, boycotts against land dispossession and resettlement of landowners in Kariakoo and Gerezani led to delays of several years (Mfinanga and Madinda 2016; Rizzo 2014). However, these boycotts were not directed at the DART project per se – in contrast to the resistance of Dar es Salaam's transport operators, who did not comply with the conditions of the World Bank to run DART under an international PPP (public-private partnership). Hence, it was not angry minibus drivers nor concerned citizens who took action, but instead a local transport company, UDA-RT, founded only in 2015 and representing the interests of a number of powerful politicians and the minibus company Shirika la Usafiri Dar es Salaam (UDA). Rather than direct and open confrontation, UDA-RT's strategies of resistance are best conceived of as a form of 'non-compliance'. Strategies included: material resistance (acquiring many buses and an advanced automated fare collection system); spatial resistance (taking over central sites of DART's infrastructure); legislative measures (obstructing the international tender through court injunctions and not complying with contracts and other forms of agreement); 'nationalistic' attitudes (as a World Bank consultant described them); and playing for time (e.g. creating time pressure in the course of the presidential elections). In short, the company did not challenge the 'if' of DART, i.e. its general existence, but the 'how', i.e. the way that DART should come into being.

Regarding UDA-RT's refusal to comply with signed contracts and other agreements, DART's stakeholders were highly indignant: 'I was never thinking of, never imagined that there could be a violation of the contract. I have never seen it anywhere else', said a transport consultant from India (AM 09/2015). Another specialist from India added: 'In India, we follow the rules of World Bank because they are the main funders of these projects, so we need to follow them' (RP 05/2016). Likewise, James Kombe from the World Bank stressed the liability of the bank's conditions: 'We need to help them [UDA-RT] to understand that there are principles that need to be followed' (JK 09/2016). UDA-RT's resistance emerged in continuous negotiations and uncertainties, which led to several amendments of the operational design in Dar es Salaam. UDA-RT successfully reformulated power relations through this strategy of non-compliance. The most challenging period of DART has thus neither been the mobilisation of the model nor the concrete planning, designing and construction. Instead, implementing the operations embodied perpetual controversies that have created a ubiquitous status of mutation in Dar es Salaam itself.

The dream of an international PPP

International organisations and companies (in collaboration with local authorities and consultants) have been determining transport planning and financing in the Tanzanian metropolis for the last two decades. The urban transport master plan reflects the narrative of global consultants: 'Transport is an entitlement to the citizens of Dar es Salaam, and good transport networks have multiplicity of benefits; socially, economically, environmentally and culturally' (JICA 2008: 14). A core element of these benefits is a new economic model based on the market-driven procurement model of PPPs (see Siemiatycki 2013). PPP frameworks have been central to numerous

World Bank projects, including BRT projects worldwide and infrastructure projects in Tanzania (Loxley 2013; see also World Bank 2016, 2017). At the East and Central Africa Roads and Rail Infrastructure Summit held 2016 in Dar es Salaam, global consultants and politicians presented PPP as the (one and only) solution to make transport more efficient, foster economic growth and reduce poverty. Even if the World Bank claims not to follow a certain ideology, it attaches conditions to its granting of credit based on bank's principles and rules. The bank welcomes PPP because it has limited trust in governments and state-owned enterprises. Consequently, the World Bank tied their financing for BRT projects, including DART, to the condition of including private sector involvement in the provision of bus services.

Persuaded that private-sector engagement is necessary for economic growth, the government of Tanzania has jumped on the PPP bandwagon. The state now tends to finance the construction of the physical infrastructure with loans while outsourcing operations to the private sector. In fact, DART was the first comprehensive PPP project in Tanzania. According to Tanzanian law, PPP projects should follow the principles of (economic) efficiency, effectiveness, and social and environmental equity (RM 03/2015). Employees of the Tanzanian PPP Unit, which provides training courses on PPP, assistance of tenders and risk management for the government of Tanzania, have quickly adopted the aim of this business model: 'It's for competition, the PPPs' (TP 05/2016). However, employees still lack practical experience, and international consultants question the true willingness of the Tanzanian government (JB 10/2015). But generally, the PPP narrative fits to the government's plan of achieving the status of a middle-income country by 2025, of which a core element is the development of high-quality infrastructure. The World Bank supports this aim and sees transport infrastructures like BRT as central to the success of this plan. As one transport consultant of the bank pointed out, transport does not solve problems by itself, but it facilitates their solution (JK 10/2015, 09/2016). At the transport summit in Dar es Salaam, James Kombe from World Bank Tanzania elaborated the necessity of government financial support in order to attract private investment:

> For you [East African governments] to have a sustainable project, you need private sector participation. […] It is a business actually, for the PPP. So that's why you need to do a lot of analysis to see how much public sector investment can help to attract the private sector participation in a project. For the BRT for example, when you look at all the business analysis and plans, which were done since 2007/8: they showed that you need a hundred per cent financing of infrastructure by the government in order to facilitate operations by the private sector.
>
> (JK 09/2016)

At the same summit, the Tanzanian Minister of Works, Transport and Communication expressed the government's openness to private foreign investment: 'We, the government alone can't do that. [...] With your engage-ment, we can make a big difference in this country', and the manager of the Tanzanian Roads Agency proclaimed: 'Designs are ready, feasibility is given, projects are ready to be invested in'. This enthusiasm for PPP was also evident as the DART project progressed. Even though PPP was quite a new concept, Tanzanian government officials had already internalised PPP ideals by repetitively speaking of 'value for money', 'enhancing economies of scale', 'facilitating foreign investment' – and they adopted the assumption that competition increases revenue and quality of service (GV 09/2016; RL 05/2016).

BRT and PPP

Across the world, PPPs are becoming an increasingly popular way of fund-ing and operating urban transport infrastructure. They are supposed to raise capital investment, transfer risks, support economic growth and sup-port sustainable development (Siemiatycki 2011, 2013). In Europe, PPPs have been associated with the increasing privatisation of public assets since the early 2000s. However, collaborations between the public and private sec-tor regarding the provision of transport services are nothing new in Dar es Salaam – even though the label 'PPP' concomitant with the terminology only came to the city with DART. The city's public transport sector had already been largely privatised in the 1980s; since then, the private daladala sector has expanded rapidly (Kanyama et al. 2004; Rizzo 2001). In the early 2000s, the Surface and Marine Transport Regulatory Authority (SUMATRA) was tasked with regulating the public transport sector, including daladala, but the actual business of daladala operations and employment has stayed in the private sector (see also Chapter 1). In addition to the annual licence fee that daladala owners pay for each vehicle to SUMATRA, bus operators pay an access fee for using daladala terminals. The daladala system also uses outsourced contractors, e.g. for security and sanitation at the termi-nals. Hence, the Tanzanian minibus system is also kind of a PPP: the system provides a basic public service at its own economic risk, but it is regulated by the government.

Rather than handing over state-run transport services to the private sector, DART's PPP arrangements have re-engaged the state in the pub-lic transport sector of Dar es Salaam. Thereby, PPPs have played a role in recent transformations of the transport sector that challenges classic cri-tiques of neoliberal configurations (see Brenner et al. 2010; Peck and Tickell 2002). Such critiques – pointing towards the disadvantages when private actors take over responsibilities of the state – have mainly been based on experiences from the Global North, and assumed an active welfare state that serves and regulates public needs (see Ferguson 2009; Ong 2006; Par-nell and Robinson 2012). In contrast, implementing BRT in cities of the Global South may actually challenge these neoliberal critiques because

BRT projects bring the state back into the transport sector (Paget-Seekings 2015). Hence, the combination of re-engaging the state through regulation and introducing transport models into the Tanzanian transport sector – that request international tenders and a direct competition between operators – brings a new kind of PPP to Dar es Salaam. This PPP arrangement only differs marginally from how the daladala system is structured: public and private operators collaborating to deliver essential services to the public is not a novelty in the city. Nevertheless, the legal frameworks of daladala and DART differ tremendously.

PPPs for BRT systems have been fiercely debated in politics as well as academia, with participants falling into two main camps. On the one hand, global BRT technocrats agree that a PPP arrangement is the best option for constructing and operating BRT systems (LL 11/2017; TG 12/2016). On the other hand, social scientists have been more critical, pointing out the downsides of turning over public infrastructures to profit-oriented actors:

> [...] the bargaining power of the public regulatory body *vis-à-vis* the private tenders is low, as is often the case with Public-Private Partnerships. Furthermore, the inclusion of previous public transport operators has often proved problematic. In Bogotá, ownership of BRT buses increasingly became concentrated in the hands of a few private operators [...], while other contexts presented their own distinctive, and at times violent, tensions over participatory inclusion.
>
> (Rizzo and Dave 2018: 2, emphasis in original)

Regarding the Global South, scholars have argued that private sector actors increasingly disadvantage public interests through the promise of development (Birch and Siemiatycki 2016; Mawdsley 2017):

> PPPs are an expensive and inefficient way of financing infrastructure and divert government spending away from other public services. They conceal public borrowing, while providing long-term state guarantees for profits to private companies.
>
> (Hall 2015: 5)

Nevertheless, protagonists from the development sector continue to introduce PPPs into diverse projects. When the World Bank's President visited Dar es Salaam in 2017, he emphasised that PPPs were not the privatisation of the 'bad old days of liberalisation'; rather, recent PPP arrangements focus on government priorities and benefits for the people:

> Government leadership and public spending will always be critical, but we've also learned that if the use of public resource is strategically focused, we can dramatically increase finance for development by crowding in private capital. [...] So at the World Bank Group,

everywhere in the world, we're increasingly asking, how can we use public resources to bring in the private sector to create a win-win for the benefit of citizens?

(Kim 2017)

However, most PPPs neither bring the promised economic growth nor reduce debt deriving from investments; instead, debts are transformed into leases that governments pay to private enterprises (Rizzo and Dave 2018). In contrast to Paget-Seekings (2015), Rizzo (2014, 2017) argues that the prevalent PPP model of BRT promotes neoliberalism in urban public transport. In the case of DART, the Tanzanian government is waiting for the system to generate enough revenue to repay the loan it received from the World Bank. DART's business model tries to maintain a balance between generating profit and operating on affordable fares, without the support of subsidies. Because the system is supposed to be a public good, it must offer services that are not necessarily profitable, for instance off-peak operations. Since PPPs are most commonly less lucrative for governments than initially advertised and expected, the win-win narrative of DART is questionable.

Opening the Tanzanian transport market internationally challenges the financial security of certain local transport actors: individual, weakly connected daladala owners, drivers and conductors, as well as the employees of UDA. The company UDA has long benefitted from its special role in post-socialist Tanzania, supported by a political elite and the Simon Group, one of the biggest Tanzanian companies.[2] There is a clear distinction between UDA and daladala businesses in Dar es Salaam: whereas UDA used to be a public company, the daladala sector is highly diverse and has less impact on politics. Due to its historical roots in the city, UDA is determined to retain its legal status as a public company, a status that brings financial benefits, even though a private company has been the major shareholder since 2011 (Mfinanga and Madinda 2016; Rizzo 2013). UDA feared losing these benefits under the framework of DART and thus mobilised its support network to disrupt the international PPP model.

According to the World Bank's funding conditions, DART was supposed to internationalise the city's service providers: the project must include in its PPP arrangement a service provider that is experienced with BRT operations – an expertise that no Tanzanian company possessed (AH 03/2015; JB 10/2015; JK 10/2015). This condition requiring collaboration with a foreign company, rather than the privatisation of public infrastructure per se, was the element of DART's PPP that led to operational controversies. While experienced international companies should ensure the quality of service, local operators can participate in operations, as James Kombe from World Bank Tanzania elaborated: 'The system itself is great, if it's done properly, it's really fantastic for the city. But it needs to run well. [...] And that requires a bit of experience' (JB 10/2015). The PPP model that most BRT consultants support recommends that private contractors provide bus operations (including fare collection and fund management), and that the state monitors, coordinates and indirectly subsidises the infrastructure, e.g. through maintaining roads. For a certain period, operators pay the government for using the public infrastructure like roads and stations. Until the DART model was created, PPP frameworks of BRT were carried out between national companies and the state.

DART's Phase 1 was planned to be the first BRT worldwide that used an international bidding process for service operations – a challenging task, as the DART Agency's chief technical advisor explained:

> Nobody had done this, a real international bid, and no financing from the government. Everywhere there's government. [...] Actually, we had illusions. We thought we could do something that has not been tried anywhere else. [...] This would have needed a very clear and almost perfect situation.
>
> (HM 10/2015)

In a similar vein, the CEO of the DART Agency Gabriel Vissanji cautioned his East African colleagues: 'There are many challenges coming up. So, the only way forward is to be prepared and cooperate between the public and private sector'. Despite these anticipated challenges, the new best practice for BRT operations – a collaboration between internationally experienced and local transport operators – was to be prototyped in Dar es Salaam. This decision was made on the assumption that the combined expertise would raise the BRT service quality. This plan was supposed to go beyond even best practices from Latin America at the time – a very ambitious plan for protectionist Tanzania, as the technical advisor admitted to me when the project had begun to totter (HM 10/2015).

BRT operations can be run according to either single or multiple service provider models, each of which pursues different ideals. The single service provider (single SP) model is more state-oriented and follows European ideals of public transport provision. By contrast, the multiple service provider (multiple SP) model is more market-oriented because it includes at least two competing operators, which can be either local or international. ITDP's preference for the multiple SP model originates from the belief that competition leads to higher service quality. Moreover, BRT consultants, citing evidence from Bogotá, Curitiba, Santiago de Chile and other Latin American cities, claim multiple SP is perfect for cities of the Global South that have few resources to directly subsidise public transport (JM 03/2015; PH 03/2015; see also ITDP 2017: 546; Muñoz and Molina 2009). This multiple SP model has won through in practice, and BRT has become a famous example of the increasing prominence of PPPs and competitive tendering models in the Global South (Hook 2005: 8ff.; Schalekamp et al. 2016: 111). In this model, the government oversees the system, handles general planning and monitoring activities and sets service level benchmarks for the private operators. Accordingly, DART's PPP arrangement allows for the government to intervene in the last instance, i.e. to bail out the project if it is in danger of failing. According to DART's consultants, this governmental backup is necessary because no private company would be willing to carry the risk alone:

> This is the game: we have to make everything possible that we're making sufficient money that the government doesn't have to subsidise this. But if there is still, the government, the Ministry of Finance to get the deal sealed, they have to sign that they will come in, to cover the deficit. Because no international investor will come as they are assured, they got the money they owed.
>
> (HM 03/2015)

PPP contracts can contain incentives and penalties to secure good performance of the private operators. One of DART's global engineers stressed the importance to incorporate local transport actors – under certain conditions:

> It's good that the local operators do the business. Because if not, you take part of the market for someone else. So, it's important that the local operator can participate. If they are good enough to do the operation: fine. The problem is: are they good enough? It seems they are not. So, they can join with international.
>
> (LL 11/2016)

The former CEO of the DART Agency followed a comparable agenda that had been suggested to her by Lloyd Wright, one of the world's leading BRT consultants and an author of the *BRT Planning Guide*: 'He told us: "The project is only successful if you have effective participation of the existing operators"' (YK 10/2015). In order to prevent resistance, the incorporation of existing operators primarily aims to maintain local structures and employment for 'political purposes' (ITDP 2017: 522). In addition, BRT operations could benefit from incorporating local knowledge and experience. This (assumed) double benefit of combining existing transport actors with bigger private companies and investors is reflected in a table that ITDP routinely presents to its audience. It reveals that BRT corridors operated under PPP arrangements involving local operators and new private investors often reach a *Gold* score and that BRT corridors operated solely by local operators often obtain a *Bronze* or *Silver* score (according to ITDP's *BRT Standard*).

Guarantee and risk

Only after UDA-RT had strategically enforced its central permanent role within the project did BRT consultants like Herbert Meyer become sceptical about DART's PPP arrangement:

> It's not really a proper environment to run a PPP tender. Either you run a real competition and then you're willing to let those guys loose or you want to secure a role for them. But then, you need to put them out on the scope of the tendering, maybe as a mandatory partner, which again is against the law.
>
> (HM 10/2015)

DART's operational controversy is primarily based upon a fundamental contradiction: how to match a guarantee given with a competition required? On the one hand, the World Bank guaranteed local transport operators to be part of DART:

> According to the contract, which has been signed, we have indicated: integration should be mandatory [...]. If integration is not mandatory, it means you can easily push them out of the system. And we are obliged to protect these people.
>
> (YK 10/2015)

On the other hand, the World Bank requested an international bidding process. Because the process was undergoing several difficulties in 2014, the bank contracted an international transaction advisory team that provided the DART Agency with financial expertise and advice on international tenders (JB 09/2016). When reworking the operational structure and the documents of tender in 2014, the team found itself in a challenging situation, as Janvier Bellamy remembered:

> We felt there was a strong desire from the DART Agency to involve those local operators. At first, it goes back a long history, but the World Bank had always promised to those local operators that there will be a role for them. It was probably an unwise promise. The World Bank director for Tanzania made a declaration when the project was initiated, something like ten years ago, and said: 'The local operators will be involved, will have a role', et cetera. And the issue is, it was never really worked out how that would actually work. And our assignment was to do it as PPP. If you do it as PPP, you have to tender it competitively and you actually can't give a special role to local companies. So, there was a sort of unresolved issue here, which had been ignored by the World Bank and the DART Agency. Which led the situation where there was a lot of expectation from the local operator's side.
>
> (JB 10/2015)

The 'long history' referred to a change of leadership within the World Bank Tanzania. At the initial planning stage of DART, ten years before the transaction advisory team was founded, the World Bank had been open to integrating previous transport operators in DART. The then World Bank country director promised in writing that daladala operators would have some role in DART. The head of the Dar es Salaam Commuter Bus Owners Association (DARCOBOA) brought the document to every DART meeting to remind the stakeholders of this promise. But with the appointment of a new country director, the official stance of the bank changed to favour an international tender. The new director backpedalled from the bank's initial promise, stating that a proper procurement process according to Tanzanian law had always been planned (JK 09/2016; MH 09/2016; RL 10/2015). In part, this paradigm shift was due to the World Bank increasingly favouring international PPPs within global trends of increasing private-sector engagement. In addition, the bank had become aware of the increasing power of UDA and realised that this power could endanger the international SP model.

Introducing the PPP Unit at the Ministry of Finance had been a complex and protracted process in postsocialist Tanzania because the state had long been ambivalent concerning PPP frameworks. Finally, Tanzania's PPP laws demand an open and competitive bidding process, regardless of the tenderers' nationalities (GoT 2010, 2011). The *Project Information Memorandum* of DART states on that score:

> Since economic liberalisation, all sectors are generally open to foreign investment. [...] The DART Agency encourages foreign firms to team up with Tanzanian tenderers in either joint ventures or subcontracting arrangements during the tender process and the execution of the services. For the Phase 1 tenders a margin of preference might be granted to local firms or associations between local and foreign firms.
>
> (DART Agency 2014b: 46–49)

How exactly this 'margin of preference' could be realised remained vague. It was only clear that local companies were encouraged to bid, but they did not obtain a special role in the selection of the service provider (GoT 2011; see also DART Agency 2014b).

Global BRT technocrats generally support PPP projects because they have little trust in governments (particularly governments of the Global South), which they associate with corruption, weak institutional capacity, and a lack technical expertise. A consultant who has been working for decades with the World Bank on African infrastructure projects elaborated:

> The key issue is that we want to get away from government as much as possible. The government has failed over the decades to deliver public services. Even though now, they are getting a bit better and better because of more skills and more money, but still, the efficiency is not there,

the motivation for innovation is not there. So, we still feel that, as much as possible, it should be private sector.

(DN 03/2015)

Remarkably, the management of UDA-RT and its major shareholder, the Simon Group, also stated that it did not trust the national government. According to the CEO, the government was a 'burden to the people' because it would not be capable managing and financing DART if the operations were not working self-sustainingly (HB 09/2016). He claimed that UDA-RT should directly take over the whole DART project, including construction and operation of the all phases:

> It's nightmare dealing with the government. I've sat with the prime minister in meetings, I think, forty meetings. And I've sat with government officials, I think, maybe eighty. In the span of six months. This is basically the first PPP. I don't think the government itself has got a much broader vision in terms of what PPP means.
>
> (HG 05/2016)

By early 2015, the governmental PPP Unit was also willing to hand over DART's operations to the private sector. However, their motivation was to reduce the government's risk (RM 03/2015; TP/GD 05/2016).

The question of risk is expressed in negotiations centred on DART's financial model: How should revenue (and debts, if necessary) be distributed? How should fares and access fees be calculated and distributed? Should subsidies play a role (even if BRT proponents argue that BRT is the only self-funding mass transit option, i.e. operating without public subsidy)? Hence, DART's fare structure was a serious matter of concern for UDA-RT and the Tanzanian government. While the government was interested in keeping the fares as low as possible, UDA-RT needed the fares to be high enough so they would be able to generate revenue and start paying back their debts. In 2016, UDA-RT argued that it could not survive with the fares approved and suggested higher fares, which SUMATRA did not approve. The government decided not to subsidise the fares (which UDA-RT considered too low) with the argument that SUMATRA had calculated so-called approved economic fares, which included a financial benefit of 15 per cent for the operator – an appropriate number to generate revenue for UDA-RT, according to SUMATRA's corporation communication manager (JR 05/2016). Because its assignment was to provide transport to the broad public, including low- and middle-income groups that had previously travelled by daladala, DART wanted fares to be as low as possible. However, UDA-RT announced that it would request a revision of the fares, which were only half what the company had demanded (AH 09/2015; HG 05/2016). When operations started in 2016, UDA-RT was under great pressure because the company had taken big loans for the BRT equipment and for compensating daladala owners.[3] In addition, the first week of operations was free of charge for passengers – a decision made by the DART Agency on ITDP's advice – in order to attract passengers and familiarise them with BRT. Financially, this was a barely workable situation for UDA-RT, whose chief operations officer told me sarcastically at the end of the third day: 'Let's see how much loss we made again today' (LH 05/2016).

Similar continuous changes and uncertainties, opposing opinions and expectations shaped the negotiations over how the revenue should be distributed among stakeholders. Debate centred on the access fee, which was tied to DART's revenue, i.e. on the fare structure and passenger numbers. According to the business model, the fund manager receives the revenue from the fare collector and allocates it to the stakeholders: fare collector, fund manager, bus operator(s) and government (DART Agency 2014b). To ensure transparency, the fare collector and fund manager are independent companies that work on behalf of the general public, because BRT is a public infrastructure, financed by a loan from taxpayers' money. The fund manager and fare collector both receive a fixed percentage of the revenue, while the lion's share goes to the bus operator(s) and the government. How much of the revenue is allocated to the government and the bus operator is based on two contrasting parameters: price versus quality, i.e. whether operations aim to be delivered in the best quality or at the lowest price possible. These parameters depend on the access fee.

Two models for calculating the access fee – either kilometre-based or passenger-based – illustrate different ways of allocating revenue and risk. Including a guaranteed share of the profit for the private operators distinguishes a PPP from an entirely state-run infrastructure, which typically has no incentive to seek profit. If the fee is based on kilometres of service, the operator always receives the same revenue (no matter how many passengers are actually on the bus), and thus has no incentive to use as few buses as possible. The bus operator receives its fixed share according to the kilometres scheduled, and the rest of the revenue goes to the government (or the loss if the uptake of services is not high enough to generate revenue). This means that the government bears the financial risk because it does not receive a guaranteed income from BRT operations. In contrast, a flexible, passenger-based fee places the financial risk on the side of the bus operator. It encourages the operator to reduce its own costs and increase its revenue by using few buses and filling them up with passengers, thus reducing service quality. This model with a flexible access fee gives the operator more freedom to determine the quality of service: the lower the operator's cost per passenger, the more profit it makes. Hence, cities striving for a high-quality BRT system tend to choose a fixed kilometre-based access fee. In addition, global consultants claim that this kind of fee is more likely to attract international companies because the government bears the risk. By contrast, if the goal is to minimise the cost of a BRT system, stakeholders tend opt for a flexible access fee based on the actual number of tickets sold (AP 09/2016; HB 05/2016; RL 05/2016).

DART's stakeholders decided on different access fee models for the long-term service provider (SP) and the interim service provider (ISP). In 2014/2015, the plan was that the SP would operate under a fixed kilometre-based fee on the trunk corridor and a flexible passenger-based fee on the feeder routes (DART Agency 2014b: 44).[4] Agreed in the *ISP Agreement*, the ISP was expected to operate under a kilometre-based access fee. This meant that UDA-RT, as the ISP, had to provide a set volume of services, regardless of passenger numbers. During 2015/2016, the situation became more complicated because UDA-RT was gaining political and economic strength through the support of certain politicians, its massive BRT fleet, and the fact that both fund manager and fare collector were collaborating with UDA-RT instead of with the DART Agency, so that the government could not access data on revenue generated. In reaction, the government became more cautious and considered allocating a passenger-based access fee for UDA-RT during the ISP period too. The updated model left UDA-RT to bear the financial risk, specifying that no government entity 'intend[s] to make any subsidy available' (DART Agency 2016: 12–13). After more than two years of UDA-RT operating DART, the DART Agency announced that the business model for the long-term SP contracts would include a flexible passenger-based access fee (Kasumuni 2018; Nachilongo 2018). Despite widespread critique of the insufficiencies of this 'low price, low quality' (LL 11/2016) business model, it fitted well with President Magufuli's rhetoric about reducing government spending. At the official inauguration of DART Phase 1 in January 2017, he stressed that the government does not carry any risk, and claimed that UDA-RT has to 'make sure to make profit', from which future development projects can be financed: 'In my outlook, there is no definition for making a loss. So if this project makes a loss, it is your loss, not my government's loss' (Magufuli in Mwangonde 2017).

(Re)Assembling agreements

The tedious process of compiling and reworking documents for DART's setup reveals the competing interests of its stakeholders, especially regarding the operational structure. The distinct operational models considered in Dar es Salaam at different points in time include a PPP arrangement with a combination of local and international operators. The contradictory requirements of the World Bank and the transaction advisory team – which sought to guarantee local operators' participation with and without conditions – could not be resolved before stakeholders requested to finally begin operations. This situation led to continuous and cumbersome negotiations of the structure, according to Amir Hanif, the spokesperson of UDA-RT:

> Details that are very, very detailed. I mean, I can understand; this is the first time to be involved in a World Bank project. So, what I came to hear is that: this is how World Bank projects are. They involve a lot of documents. A lot of documents. A lot of discussions. You cannot

imagine, we have been discussing this since six months now. [...] For the ISP, the interim service, the interim term sheet. And then, the draft contract and what should be in and what should be out. What we can do and what we cannot. The do's and the don'ts. You see. I mean, it's a very, very complicated, it's bulky and it's complicated.

(AH 03/2015)

This controversy can be best understood by chronologically tracing the four steps of (re)assembling the legal agreements between the DART Agency and UDA-RT: from a multiple SP to single SP (step one); from SP to ISP (step two); from the *ISP Agreement* to *ISP Addendum* (step three); and from a permanent SP to single operator (step four). These four steps show how the 'detailed details' provide a record of power relations and the strategies adopted by local transport operators.

Step one

During different planning stages, both single and multiple operator models were considered for DART (DART Agency 2014b; DCC 2007). Global BRT technocrats contrasted the two models and criticised the respective other model for being either too ambitious or having poor service quality. Inspired by *Transmilenio*, the Brazilian consultant applied the multiple SP model in the initial DART design from 2007. According to this model, the government coordinates four separate components: two competing bus operators (each a collaboration of local and international operators), one fare collector and one fund manager. The DART Agency would have had to coordinate the different components and bear the financial risk (HM 03/2015; JB 10/2015). By January 2014, the design had become out-of-date due to delays in implementing the system. The transaction advisory team conducted a market analysis and reviewed the business plan and financial model. The team's main goal was to reduce the risk for the government, as Bellamy from the transaction advisory team explained:

> Our thinking was: giving the current capacity of DART [Agency] – the fact that they don't have experience yet, it's a new project starting from scratch – we thought that a lot could go wrong in this process. [...] You end up in a situation where government has to pay whereas actually the service is not there.
>
> (JB 10/2015)

The transaction advisor shifted the operational risk as well as the communication and coordination between the different operators to the private sector, and reduced the SP model to two separate contracts: one communal contract called the 'single SP', which included a fare collector and two bus operators (with both Tanzanian and internationally experienced companies in collaboration), and one contract for an independent fund manager (to guarantee transparency about the revenue distribution). Including a local operator was encouraged but not obligatory, as the *Project Information Memorandum* indicates:

> While it is widely understood by current daladala owners that they will not enjoy preferential treatment in the tender process, the inclusion of parties with local experience within bidding consortia will be encouraged following the possibilities offered by the PPP and procurement legislation and regulations of Tanzania.
>
> (DART Agency 2014b: 39)

In June that year, the transaction advisory team presented the single SP design to interested bidders at a market consultation meeting (DART Agency 2014c). Whereas the Tanzanian government had no uniform opinion about this newly proposed arrangement, and local operators were mainly concerned that they would not have a chance to compete with international bidders, the chief technical advisor of the DART Agency distanced himself from the multiple SP model:

> I understood: this [multiple SP] would be a disaster. [...] We looked at Bogotá. You know, they were the first ones who did this. And they also learned. But that was ten years ago. Things develop, you know. So that was the wrong thing, I think, we took over from them.
>
> (HM 03/2015)

Three months later, the Prime Minister himself decided for the single SP model.

In the meantime, the biggest local transport actors in Dar es Salaam, UDA and DARCOBOA, realised that they had to strengthen their position to have a chance operating DART. Together, they founded UDA-RT. Bellamy commented:

> If I were UDA, I would have done exactly the same. I would make an alliance with them [DARCOBOA] so that in front of government, they represent all of the local operators' interest. [...] UDA-RT signed sort of an association agreement saying that they would be together for the SP tender, but they would not associate with any international company. This doesn't make sense because they can't provide all services, like the fare collection services.
>
> (JB 10/2015)

The managing director of UDA-RT management explained this step:

> The World Bank and the transaction advisor wanted to use their advantage of the high fragmentation of the daladala operators, to bring

in one big external operator, a foreign operator. [...] They just left them like that, fragmented as they were so that they can be denied any opportunity. Because the way the daladala operators are, first they don't have institutional capacity. They don't have financial capacity, they don't have technology and all that. They are just fragmented. One by one, one by one. How could they come in to a bid for this cooperation? This is my question. It was not easy for anyone to come to bid. That's why we have to come together so that we are competitive.

(NA 09/2016)

For the single SP model, at least two competent Tanzanian operators were needed for the two collaborations of Tanzanian and international bus operators. Hence, founding UDA-RT made it pointless to tender for another Tanzanian operator; it 'killed the deal' since only one capable Tanzanian partner was left to participate in an international deal (HM 03/2015). This strategy temporarily removed DART from the market.

Step two

In September 2014, the Prime Minister not only opted for the single SP suggested by the transaction advisory team; he also introduced the ISP. This decision was in response to ongoing delays in the SP procurement process (GV 09/2016; PF 20/2015). The government estimated that procuring the long-term single SP would take two years: one year for procurement of the service providers, and one year to purchase buses and install the automated fare collection system (AFCS). The ISP was expected to run operations in the meantime (DART Agency 2015a). According to the DART Agency (2018), the decision to use an ISP was designed to provide a training ground for the long-term operators, while also providing relief to Dar es Salaam's commuters, and utilising the physical infrastructure so that it was less likely to be misused or vandalised. Another goal was to give local operators the opportunity to gain practical experience with BRT so that they could increase their ability to collaborate with an international operator. The ISP concession was handed to UDA-RT without a tendering process. After years of delays, each month counted, especially with presidential elections approaching, in which the then Minister of Works, John Magufuli, was running (see Chapter 4). The government expected that the ISP would start operations in September 2015, one month prior to the presidential elections.

The conditions of the ISP are stipulated in the *ISP Agreement*. As Amir Hanif explained to me, it took UDA-RT and the DART Agency six months to negotiate the agreement, finally signing it in April 2015. In the agreement, the transaction advisory team had given the DART Agency the power to assess if UDA-RT were acting according to the conditions, and if UDA-RT was adhering to the requirements of a long-term SP in an international PPP arrangement. This condition implied that UDA-RT would have to bear the cost of buying buses and a ticketing system itself. The transaction advisor sought to guarantee that a long-term SP will still be tendered in which UDA-RT has to be incorporated – under the condition that UDA-RT's quality of service correspond to the standards set (JB 09/2016; see also DART Agency 2016). The *ISP Agreement* further stipulated that UDA-RT would operate DART Phase 1 under a single-source contract with a simplified ticketing system and reduced number of buses. The transaction advisory team calculated that the ISP should operate with 76 buses, while the long-term SP would have 305 buses. Moreover, UDA-RT was to compensate relocated daladala operators and pay a fee to the government for use of the physical infrastructure (HB 09/2016). This arrangement was globally unique, involving novel plans, practices and standards. Until then, no other system had used an ISP, nor was this option considered in the global BRT model. This might be due to its various difficulties, which have been particularly criticised by the World Bank.

Despite the care taken to formulate the agreement, the contract was suddenly changed the day before it was signed. While the transaction advisory team and the World Bank were absent, the DART Agency (under pressure from higher levels of the Tanzanian government) altered the contract:

> [The changes added] basically more flexibility for ISP to increase the scope of services and securing a place for them after the interim period. We said in the initial *ISP Agreement* that they would have to agree to a reasonable integration approach with the future PPP preferred party. And if they did not agree to something reasonable, then the government could still push them out. And this was removed. So basically, it meant: whatever the regime of the PPP tender was, there would be a role for ISP. So, they were basically surrendering all the decision power to UDA, and UDA could basically dictate the condition under which they would stay because the agreement says: 'They will stay.'
>
> (JB 10/2015)

With this last-minute change, the DART Agency lost its leverage. The signing ceremony of the agreement took place a day later. Witnessed by several high-ranking government officials, it seemed to be a relief for everyone involved. The DART Agency ran an article on its website titled 'High Hope for BRT Project as Key Players Agree', reporting that UDA-RT's spokesperson 'pledged to offer good services' (DART Agency 2015b). Subsequently, instead of simply encouraging international companies to incorporate local operators, the DART Agency merely stipulated that they should express willingness:

> DART Agency launches an international competitive bid and will select someone with BRT and Smart card experience. To be selected such bidder must show willingness to integrate with the Interim Service Provider.
>
> (DART Agency 2018)

Step three

The World Bank fundamentally disagreed with using an ISP because the single-source contract did not follow a proper PPP structure. Consultants saw the risk that UDA-RT would try to take over the whole DART project (JB 10/2015; JK 05/2016; PH 07/2015), a fear that proved to be well-founded, as the head of the transaction advisory team expressed:

> Obviously, as soon as you sign the *ISP Agreement*, there is a strong in-centive for ISP to try and grow the scope so that there is absolutely no room left to run a tender thereafter. And then, they're running the whole show. And that's exactly what is happening.
>
> (JB 10/2015)

UDA-RT breached the *ISP Agreement* by making a significantly larger in-vestment than agreed upon, purchasing almost double the stipulated num-ber of buses as well as a complex AFCS technology instead of a simple ticketing system. Even if research partners guessed that some government officials must have known about this plan ahead of time, UDA-RT's strat-egy to resist the contract only became public when 138 buses arrived at Dar es Salaam port in September 2015.[5] Thereby, the company succeeded in ex-tending its material and operational scope – and consequently its position within DART. As the CEO of UDA-RT stressed:

> This is a huge investment. It's actually not something, which you can throw away. Therefore, the way I see the future is: either it's an integra-tion with the service provider or an extension. I actually see more on the extension side.
>
> (HG 05/2016)

Global consultants were not surprised by UDA-RT's strategy. Indeed, the transaction advisory team had warned the Tanzanian government that an *ISP Agreement* would encourage UDA-RT to increase its scope so that it became indispensable, which was contrary to the notion of an international bidding process: 'Because it's interim, the situation is not well defined. This is not the way to operate because you need to impose some rules' (LL 11/2016).

This non-compliance with the *ISP Agreement* resulted in the *ISP Addendum*. The government could not just cancel the deal with UDA-RT, even if it had wanted to. Both sides heavily depended on the other: the government needed a bus operator, and had no alternative to UDA-RT; and UDA-RT needed to utilise the massive investment it had made in order to repay its huge debts.

This breach of contract not only led to UDA-RT repositioning itself within the DART project, it also impacted on the process of modifying the long-term SP model. As a result of that investment, it became clear that the government had to ensure UDA-RT's long-term integration in DART's operations. And because UDA-RT made its move right before the presidential elections, the decision regarding the long-term SP was postponed until a new government was in place. As Bellamy elaborated shortly before the elections:

> Anything can happen. There's a lot of issues about corruption in Tanzania and the way things are being done is usually: you can send your political enemies in front of a court for some corruption case. So maybe the next party or president has a very different view and says: 'This deal was all corrupt.' I don't know; this can solve the whole interim service deal. It's also a possibility. I don't think it's very likely.
>
> (JB 10/2015)

Developing the *ISP Addendum* took another several months, and involved a new negotiation of fares and fees. Due to the amendments to the scope of services supplied, stakeholders requested to increase the infrastructure usage fee (as UDA-RT would use the government's infrastructure more intensively with the higher number of buses), while UDA-RT found the fee too high and requested it be reduced. After a public hearing in January 2016, the Tanzanian court moreover decided that UDA-RT had to clear the buses and AFCS equipment with the Tanzania Revenue Authority (TRA) – a process which, in conjunction with Magufuli's anti-corruption politics, allowed the national government to regain some leverage over UDA-RT (see Chapter 4). While the two prototype buses had already been registered for training purposes in August 2015, the rest of the buses could only be registered and used on public roads after they had been fully cleared (see Figure 5.1). The World Bank finally advised the government to make use of the BRT equipment

Figure 5.1 DART buses at Kivukoni Terminal (author's photo, 09/2015).

that UDA-RT had brought in. Just a few days before the inception of interim services in May 2016, the *ISP Addendum* was signed, and the new fare structure was approved (AH 10/2015; JR 05/2016).

The *ISP Addendum* was kept under wraps. Research partners from UDA-RT and the government were rather suspicious of my interest in this 'unspectacular' three- to four-page long document. They told me that the document basically says that UDA-RT is one of the two long-term bus operators of DART, and that UDA-RT is allowed to use all buses and the AFCS equipment. The AFCS will be handed over to the DART Agency after UDA-RT's conversion from the ISP to the SP so that the government can regain control over operations and oversee the distribution of the revenue (TP/GP 05/2016). Hence, from the initial level of 'encouragement' (*Project Information Memorandum*) to the median level of 'willingness and intention' (*ISP Agreement*), UDA-RT reached the level of 'expected extension upon satisfactory performance' (*ISP Addendum*, c.f. DART Agency 2016). This operational upgrade and power increase of UDA-RT represents a conversion from ISP to SP.

UDA-RT further manifested its position through its spatial practices, which brought a new dimension to the controversy and had political consequences. The *ISP Agreement* declared that the 76 buses should be parked overnight at Kivukoni Terminal and Gerezani Terminal. According to the *ISP Agreement*, the bus operators were supposed to have access to only a few, smaller offices in the side wing of Jangwani Depot, while the DART Agency was to share the main office with the fare collector. However, after signing the *IPS Addendum* in May 2016, UDA-RT was allowed to move to Jangwani Depot, which included the bus depot, workshops, offices and the control centre. Acquisitioning the AFCS made UDA-RT the principal entity for collecting fare and overseeing operations together with the fare collector, and the 140 buses they had ordered did not fit anywhere else (HG 05/2016; HM 10/2015). Stakeholders had a variety of perspectives on these developments: while DART Agency staff and international consultants saw them as a takeover, UDA-RT believed it was its right as the operator to be located at the heart of the system.

There is no way of knowing whether the two main decisions – for a single SP and for UDA-RT to act as ISP – were inextricably linked. The government's course of action is questionable, given the last-minute changes and the fact that influential government officials were closely connected to UDA. Following rumours and allegations that international stakeholders were involved in the Tanzanian transport sector, the DART Agency had acted on instructions from high government levels, which are deeply intermingled and influenced by the private transport sector. The porous boundary between the public and private sector indeed enabled the decision on the ISP to be pushed from the top government levels down, as James Kombe from Word Bank Tanzania noted:

> You have a very weak government in Tanzania in general, which makes the option of going for the full competition tender very difficult. The

government can easily be influenced by the private sector, by people like UDA et cetera. So, it's not the best environment to run a competitive tender.

(JB 10/2015)

The events have shown that UDA-RT found its way into DART as permanent bus operator via the single-source ISP. The promise of being converted from ISP into SP guaranteed UDA-RT a long-term role in DART's operations, as Amir Hanif from UDA explained:

> Our agreement with the government is that we will be part and parcel of this project. Come anything. Whoever is tendering, competitive tendering, bidding, whatever. You name it. Whatever is going to happen, we must, in our agreement, we must be part of this BRT project for the next twelve years.
>
> (AH 10/2015)

Step four

The final stage of the operational controversy concerns whether the Tanzanian government will procure a second bus operator, or whether UDA-RT will manage to remain the sole bus operator in the long term. This issue was articulated in two lawsuits; UDA-RT spared no effort to prevent competition. Juma Ruete from SUMATRA complained about the company: 'They do not comply with the law at all'[6] (JR 03/2015), and the consultant Luc Lavoie emphasised that for this very reason of non-compliance, an international tender is necessary:

> It's very political. [...] That's why you need to bid. And you need other companies. [...] Then, UDA-RT can adjust. They need to respect the rules when there is another company involved.
>
> (LL 11/2016)

After the DART Agency finally managed to initiate an international bidding procedure for a second bus operator in June 2016, UDA-RT successfully objected that it should also be allowed to bid (JB 09/2016). As Gabriel Vissanji from the DART Agency elaborated:

> It restricted him [UDA-RT] from bidding when we came to open tendering because he was already a service provider. He was only to be integrated with the winner. So, he was against, he also wanted to be a bidder. So, now we have corrected that he can also bid as a separate type.
>
> (GV 09/2016)

UDA-RT's self-confidence had increased after starting operations. The managing director of UDA and the Simon Group described the situation as risky, but added that it would be a 'very unlikely and unpleasant step' of the government to exclude UDA-RT from the long-term DART project (NM 05/2016). In October 2016, the UDA management stated:

> It is clear that we are allowed to bid, as ISP. And if we win, then we will be single service provider. If we lose, we will be converted into SP. So, there will be two operators. So, in the other way, we will be awarded a longer period. [...] ISP contract says that we will be integrated with the SP.
>
> (HB 09/2016)

A few months later, UDA-RT changed its mind and tried to annul the government's attempt to tender for an additional bus operator. The company argued that this tender would breach the *ISP Agreement*, which guaranteed exclusive rights for UDA-RT. According to UDA-RT, bus operations would not be commercially viable for two bus operators because supply would then exceed demand. However, the government won this second lawsuit and the DART Agency re-advertised the tender in May 2017 (Nachilongo 2017; The Citizen Reporter 2017). Despite losing the lawsuit, this dispute brought UDA-RT almost one year of extension of the official ISP period because the international tender had been put on hold until the Tanzanian court reached a decision. While preparing for its *Mobilize* summit in Dar es Salaam in June 2018, ITDP pushed for a decision to be made beforehand (ITDP 2018b; Nachilongo 2018). However, two years later – four years after the ISP began operations – still no second bus operator has been selected.[7]

Material resistance

UDA-RT not only made last-minute alterations by changing the content of the *ISP Agreement* right before signing, it also altered the *ISP Agreement* through material practice after the agreement had been signed. Objects like the buses and the fare collection equipment do not exist in isolation; rather, they evolve through practice. The company's strategy has had sociomaterial outcomes, which are the result of controversies and negotiations on agency and power. James Kombe from World Bank Tanzania said:

> We helped them [the Tanzanian government] to damage the risk as much as much as we could by limiting the scope of what the ISP should provide in the ISP Agreement – in terms of limiting the number of buses, of limiting the involvement of the fare collection equipment. So that there would still be something to tender as a PPP thereafter.
>
> (JK 10/2015)

This attempt to limit the scope of UDA-RT failed because the company used exactly those two components – the buses and fare collection technology – to cement its role as permanent bus operator: UDA-RT's non-compliance to the *ISP Agreement* was articulated in the form of 56 extra buses and an *advanced* electronic ticketing system. Thus, controversies materialise in specific forms that have concrete consequences, such as changing power structures. Material objects can transmit or 'carry' facts (Valeriani 2010: 43) so that political negotiation happens through the material since the material is central to the social (see Latour 1991, 2005; Law 2004). Purchasing those two technological components affected the overall structure of the *ISP Agreement* and the *ISP Addendum*. The strategy secured UDA-RT more power because these components are of high financial and symbolic importance for the DART project. Material resistance turned out to be an effective form of non-compliance that was followed by a number of ensuing debates and changes.

The buses, from a Chinese manufacturer, and the AFCS, from a Belgian supplier, created new precedents and relations. With the arrival of these technical components, the controversy between DART's stakeholders reached its peak. When buses arrived in the port, it became obvious: with this purchase, UDA-RT made itself irreplaceable and would operate DART for a long time. The material became political, and resistance acted out materially. As the advisor Herbert Meyer declared, some of DART's stakeholders became increasingly gloomy about DART's future as an international PPP project:

> It's now obvious, that they are the SP. They are no longer the ISP. And there's no reason to, there's not going to be another operator [...] You can't run a system that way. It has been messed up for an international bid; no one will be interested. That's the failure of the world class system.
>
> (HM 09/2015)

According to Meyer, DART was no longer attractive to international bidders, because they would now be forced to cooperate with UDA-RT and the financial risk would be too high. Other Tanzanian government officials, however, were not surprised by UDA-RT's strategy: 'UDA-RT took the opportunity as business men, that's normal'[8] (RL 10/2015). Juma Ruete from SUMATRA commented: 'Interim will end up permanent if you are not careful. Because the guys will keep on buying more cars' (JR 10/2015).

Critics were also suspicious that relevant government officials had known about – and possibly even supported – UDA-RT's plan. The DART Agency has 'unfortunately been captured by politics', as an external consultant told me a month after the buses had arrived in Dar es Salaam. Legally, the Tanzanian government could have cancelled the contract because conditions had not been met and deviations of the plan were only permitted 'after written approval from the Agency' (DART Agency 2015a). But this was not to happen, not only for political reasons but also for pragmatic ones, as another international consultant elaborated:

> They [UDA-RT] know they will stay. After they began, you will not say: 'Ok, goodbye'. You know, after two years. The guy will never get

his money. So, you need to understand the operator also. [...] Now, you need to incorporate the thing. UDA-RT will be there. But then, we need to change the structure, improve, get some other operator. Change those things. But they will be there. This is a fact.

(LL 11/2016)

The higher number of buses and the advanced AFCS became a financial risk not only for UDA-RT, but also for the government of Tanzania. These two parties had to negotiate the clearance, and to renegotiate the fares, access fee and schedules according to the actual dimension of UDA-RT.

Like the contents of the mysterious, allegedly unspectacular *ISP Addendum*, the actual costs of UDA-RT's considerable investment in the extra buses and the AFCS, including comparably high Tanzanian import duties and after sales services (i.e. training staff on operations and maintaining buses and AFCS), were not published. Even if UDA-RT signed the *ISP Agreement* in April 2015 – and thus agreed to the limited scope – the company's management argued after the purchase: 'That was the worst contract I have seen in my life. [...] Because you are buying an asset of eight to ten years for a contract of two years. It doesn't make sense' (HG 09/2016). Amir Hanif from UDA-RT explained:

The fact is that when you order buses, the more buses you order, the cheaper. [...] The government decided to have the ISP between Kimara and Kivukoni, which was last year. But when we were ordering these buses, the other portion of the corridor, Msimbazi and Kawawa Road, it's almost [done] by eighty per cent. So, we took the risk as business people [...] and we said: 'Why not taking more buses? They might be needed'. It's rather having them here now, rather than coming next year in January, when the government says: 'Oh if only you can bring some more forty, fifty buses for this stretch'. So, this is the second reason. And the third one: It's just estimates. Who decided 76?

(AH 10/2015)

Hanif proved to be right. The DART Agency's director of operations and infrastructure management even assessed the 56 extra buses as a 'blessing' (RL 10/2015), and agreed that they could be used for the newly completed stretches of Msimbazi Street and Kawawa Road, which had not been considered in the *ISP Agreement*.

The purchased AFCS (see Figure 5.2) was not a 'simple electronic ticketing system' (DART Agency 2015a): it issues and accepts contactless media like smart cards, does routing and scheduling, delivers public information in buses and stations, and coordinates the fare collection between the stations and the bank (AE 05/2016; GV 09/2016). UDA-RT considered this advanced electronic ticketing system to be necessary because it had not been able to find a 'simple e-ticketing' solution, even though the company had

Figure 5.2 Turnstiles with smart card readers at a DART station (author's photo 05/2016).

searched the world for it (AH 10/2015). By acting alone, UDA-RT excluded the DART Agency from the decision on what kind of AFCS would be best for DART in terms of its expandability in future phases (RL 09/2016). Since UDA-RT was the only stakeholder with access to the data processing, the DART Agency complained that UDA-RT had arrogated the financial and operational planning to itself. Hence, data became a powerful instrument in DART's controversies: whoever had access to this valuable information controlled the operations and the distribution of revenue. The DART Agency planned to oblige UDA-RT to hand over the AFCS and the control centre to the DART Agency so that the government would regain control of DART's operations and revenue. The DART Agency emphasised that the government had to guarantee transparency of operations and spending of revenue because DART was financed with public money. During the first years of the ISP operations, UDA-RT and the Tanzanian government made contradictory statements on who could access the data. The service delivery manager and the chief operations officer of UDA-RT claimed that the company ensured transparency in this project by giving the government access to all data (AH 05/2016; ZD 09/2016). At a transport conference in Nairobi, the CEO of the DART Agency contradicted UDA-RT's claim, saying that UDA-RT had not yet published fares collected and revenue generated.

Principles and power games

The assembling of DART has been complex and the relationships between the World Bank, the DART Agency and UDA-RT have been hallmarked by tensions. The national government, represented by the DART Agency, has been in a difficult position throughout the process as it tried mediate between local interests and international conditions. Moreover, it has been challenged by conflict between powerful politicians on the one hand, and between different government levels on the other hand. The government has had to constantly assert its authority against both the World Bank and UDA-RT, especially when it comes to enforcing the international PPP model and dealing with resistance against it. However, every entity depends on the others: the Tanzanian government needs the World Bank as financier for diverse infrastructure projects, including Phase 3 and 4 of DART; the DART Agency needs UDA-RT to operate DART; UDA-RT needs DART to generate revenue; and the World Bank needs DART to be successful to encourage future investment in BRT, and to have a model of African best practice for future PPP arrangements.

UDA-RT's non-compliance has however forced the government to play the local company's game. After the arrival of the buses from China and the AFCS from Belgium, the biggest leverage the Tanzanian government had was to enforce the full payment of import duties for the purchase. UDA-RT tried to circumvent this full payment, as the CEO of UDA-RT justified:

> These are public buses. And they were supposed to get in the country tax-free. But when they arrived here, TRA said: 'Pay taxes'. And we didn't expect that. It took us about four months to actually sort that one. We have paid taxes with a loan. [...] It is part of the political agenda. Because we tried to actually look for an exemption and we couldn't.
>
> (HG 05/2016)

This decision was highly political and (dis)pleased several central actors; it ran counter to the anti-corruption politics of the newly elected President Magufuli on the one side, and the time pressure to finally start operations on the other side. This strict decision has historical roots, which explain the government's carefulness and harshness with UDA and its sister company, UDA-RT: Back in 2012, UDA had imported three hundred buses, supposedly for future DART operations, but without official permission from the DART Agency (AH 10/2015; HB/NM 09/2016). It then claimed the buses were exempt from import duty due to UDA's status as a public company. The government disagreed but made a compromise to clear the buses in advance. UDA was supposed to pay the import duties in instalments but never paid anything (Msikula 2015; Rizzo 2014: 21). It was against this background that UDA-RT made the huge purchase in 2015, and invested in another seventy imported BRT buses in 2018. On this third occasion, the government once again contended that it had not known about the purchase until the buses arrived at Dar es Salaam port. Like in 2015/2016, the TRA refused to clear the buses until the import duties were paid (Malanga 2018; Sauwa and Elias 2018). These events show that it has become a tried and true strategy of UDA and UDA-RT to expand their fleet despite lacking an official agreement with the government. This material growth of the two companies is the means towards increasing political and economic power.

The new Magufuli government used all its leverage to ensure that UDA-RT did not succeed with its strategy. In early January 2016, Magufuli set a final deadline for the payment of the import duties and therefore the start of operations (GV 05/2016; Peter 2016). That the deadline could not be met was only announced by the government a few days beforehand. A leading employee of the DART Agency wrote me four days prior:

> We are still not sure if the buses will start on the 10th of January because there are still various issues to be resolved; like the fare collection system, and SUMATRA needs to complete the process of fare structure, et cetera.[9]

Only after the full clearance of the buses and devices in May 2016 was the fare structure finally decided (RP 05/2016). The World Bank also used its leverage to intervene. Even if consultants claim that the World Bank has no own interest – despite the 'principles to be followed' (JK 09/2016) – the bank enforced the international tender by attaching a condition for the planned funding of constructing future DART phases, as the CEO of the DART Agency expressed:

> It all depends with the Phase 1. As you know, we are currently doing procurement of the SP [international service provider], which was one of the key issues which they [the World Bank] wanted to see done, in order for them to support. So, they come to see: 'How is your procurement going on?'
>
> (GV 09/2016)

The World Bank cannot act contrary to its own ideals and contracts because it cannot risk being seen to have been bribed or outwitted. This would mean a significant loss of face, which would reduce its credibility and authority. However, the World Bank softened its stance slightly during the protracted process, reformulating its condition for Phase 1. As the bank's senior consultant James Kombe explicated:

> We don't want to live in mess while we are in Dar es Salaam. So, we also advised the government during this mission to, among other things, to amend this contract and utilise all these buses. [...] As long as the interim does not jeopardise the future operation, that's fine.
>
> (JK 10/2015)

The World Bank seemed to have two main reasons for this changing tone. First, the bank sought to tender for a second bus operator in order to maintain the idea of an international PPP, with DART as a proxy for other BRT projects in Africa. Second, the World Bank wanted to secure competition because UDA-RT had become unexpectedly strong. In our interviews, James Kombe was emphatic about the importance of bringing any other operator into DART – either another bus operator like a long-distance bus company, an international fare collection company or a company assisting UDA-RT with business or financial management (JK 05/2016).

The illusion of an international PPP

The dream of an international PPP turned out to be an illusion. UDA-RT succeeded in permanently changing DART's operational model, because (so far) no international company has been seriously interested in taking ownership of DART's operations. The Tanzanian transport company promises that its interim service is just the beginning of more services to come:

> During the second week of the ISP operations, UDA-RT publishes an advertisement, which is captioned: 'The beginning of changes. Opportunities for everyone'[10]. The picture shows cars and dala-dala lining up in the mixed-traffic lanes on Morogoro Road towards the city centre. The physical infrastructure is still under construction; the corridor is completed, but neither stations nor pedestrian walkways have been completed yet. Since DART does not operate yet, pedestrians use the median BRT lane as wide walkway. A sky-blue BRT bus with a shiny and big UDA-RT logo and a hardly visible DART logo, subsequently inserted into the image, gives an idea how this change could look.
>
> (May 2016)

UDA-RT has become indispensable to DART's material and operational structure, with its large bus fleet, the AFCS and its headquarters at the prime location, Jangwani Depot. The company has dealt with contracts and agreements in its own way, which can either be seen as non-compliance or as a breach of treaty, depending on one's perspective. Through material practice, i.e. the acquisition and implementation of BRT-relevant components, the local transport operator has been able to enforce its interests. In addition, UDA-RT managed to take over central functions of DART, such as accessing and controlling the data, and thereby reduced the government's role to a minimum. Hence, the PPP framework of DART shows significant similarities to the structuration of the daladala sector: the government regulates the system by determining fares and handing out licences to the operator(s), but it does not obtain the overall control of the system.

A major concern of policy mobilities literature is how parts from elsewhere are assembled in a specific locale (McCann 2011: 144; see also Healey 2013). Even if the role of global consultants is not to be underestimated, economic and political elites from the site of arrival ultimately decide whether to adapt or resist a policy, and how it needs to be reshaped. This means that BRT adaptation is locally driven and actively integrated into a wider system of power relations of policy circulation (see Wood 2014, 2015a). UDA-RT's material resistance was clearly locally driven, but it could not have taken place without its global connectedness. The AFCS from Belgium and buses from China are only the surface of a complex network that relates to the formation of DART and the aspect of non-compliance. The current DART operations deviate from the global BRT model and from the operational model of DART itself. UDA-RT's non-compliance blasted the hope of global consultants that DART's international SP model could easily circulate to other sites as the DART model. Herbert Meyer remembered: 'We wanted a perfect thing. We thought that with the help of the World Bank and some good governance we could get it done. And that was a complete illusion' (HM 10/2015). The technical components of BRT have become highly political and turned DART into a technopolitical controversy (c.f. Mitchell 2002).

It is worth noting that the only actors who actually raised controversies were neither experts, nor were they laypersons such as active citizens (see Callon et al. 2009). Instead, they were actors in-between, like UDA and DARCOBOA. The UDA-RT leadership is not the typical daladala sector because it has ample financial and institutional resources. The company's privileged position gives the company a self-assertive demeanour that has allowed UDA and DARCOBOA to persist throughout the process. This means that the endeavour to democratise DART's planning process by reclaiming a globalised project did not come from below but from in between. It is not active citizens that have not been included in the hybrid forum of DART, but rather an active elite.

The UDA-RT model

As Dar es Salaam's central bus operators, UDA and DARCOBA embraced one of the main arguments global BRT consultants advance at conferences: cities can make use of local structures and capacities when choosing to implement BRT. This simple but powerful strategy of UDA-RT – merging to become irreplaceable – is an emancipatory approach, which could possibly circulate globally as the 'UDA-RT model'. This model would demonstrate options to transport actors regarding how to react to an upcoming restructuring of the sector due to BRT plans: that is, resisting international operational SP models by enforcing the democratisation of planning through socio-material practices. The materialisation of UDA-RT's strategies proved to be highly effective at reopening negotiations on DART's service provider model. It is possible to read this strategy not just as a breach of contract, but as a translation of DART's formation to its context-specific needs. Presenting the government of Tanzania and the World Bank with a fait accompli was probably the only way for the local transport actor to assert its position.

In order to enforce its will, UDA-RT had no choice but to act contrary to agreements in a non-transparent and strategic way. The materialisation of non-compliance (in the form of additional buses and an advanced AFCS) forced a reaction, whereas merely demanding to rediscuss the model could have just been defeated. UDA-RT used unconventional means to enforce a democratic planning process, most notably by creating facts with irrevocable changes. The company needed to break the rules to exist. Pursuing the target to have a secured permanent role in the DART project, the company could not comply with certain World Bank conditions that were part of the PPP model and the *ISP Agreement*. For instance, UDA-RT management was aware before signing the *ISP Agreement* that the short timeframe of the contract, combined with the high amount of investment required, made no sense. The management did not address this aspect openly before signing the contract but rather chose the path of retrospective non-compliance. Since controversies can only be solved and lead to a stabilisation of the technopolitical system through debate, a system is only stable once it has been through an open, democratic debate by a variety of actors that decide collaboratively on the shape of the system. This is exactly what UDA-RT did: it enforced a hybrid forum in order to restructure power and rework decisions.

The political economy of Dar es Salaam's transport sector plays a crucial role in the debate regarding whether the business stays purely Tanzanian or whether an international company will be involved. International consultants criticised UDA-RT's 'Tanzanian mentality'. According to Bellamy from the transaction advisory team, Tanzanians were suspicious that foreigners just want to take capital out of their country, which makes it relatively difficult for foreign companies to become involved in Tanzanian business activities (JB 09/2016). One World Bank consultant commented on UDA-RT's vehement endeavour to become the single bus operator: 'They are too nationalistic' (JK 05/2016) – a critique that is reflected in a statement of an employee of UDA: 'We are Tanzanians, we have to offer the services'[11] (AK 05/2016). The management of DARCOBOA and UDA claimed the DART project for themselves – not only because of their nationality, but also because of their expertise in local transport operations. For instance, UDA-RT committed itself to employing international BRT specialists as a contractual part of the *ISP Agreement*. The operations specialist Rachit Padmanabhan, who had set up one of the most acknowledged BRT control centres worldwide in Ahmedabad, signed a two-year contract as operations manager. However, he left the project early due to diverging understandings on how to operate a BRT system, which allowed UDA-RT to employ two Tanzanian transport specialists instead.

Amir Hanif and Harold Bafadhili from UDA-RT leadership compared DART's international PPP arrangement with a forced marriage:

> One of the World Bank consultants from South America came here last year and he asked me: 'How come, this kind of agreement? I have not

come across with this agreement of the government'. And I asked him: 'Why?' He said, in their country, [...] there are some marriages, which are called shotgun marriages. Shotgun marriages. So, they come with a gun and they force the marriage. It's kind of a forced marriage. [...] Now, where are we standing now? Before, you [the World Bank] said we must be part of it. Now you're saying: 'It's a forced marriage'. Now, what is this now?

(AH 10/2015)

They [the World Bank and the Tanzanian government] just make out each other from someone not yet born and another one. You arrange a marriage; a marriage of people who are not yet born. Saying that the preferred bidder for SP will integrate the ISP. Which is like a forced marriage. Which was considered to be probably a danger to the wellbeing of the project.

(HB 09/2016)

According to UDA-RT, the World Bank's model would not have been a win for the future of the country as a whole, and its economy in particular – because forced marriages do not turn out well.

UDA-RT saw in DART an 'opportunity for everyone' (AH 03/2015). The company was not generally against a new mode of public transport but questioned the advantages an international operator would actually bring. UDA-RT's shareholders disagreed with two main conditions of the DART project: the forced marriage of local and international transport operators, and the direct involvement of the Tanzanian government in the public transport sector. Bafadhili commented: 'We prefer single service provider because with our experience, because UDA is UDA' (HB 09/2016). According to the chairman of UDA, companies are more reliable and innovative than the government because they obtain better funding possibilities and professional experience. Since the Tanzanian government was not capable of subsidising public transport, UDA-RT needed to take responsibility and make DART profitable (HB 09/2016). According to UDA, the structure of a single SP should include not only bus operations, but also fare collection and fund management. UDA made its position clear: the company sought to own the whole DART project (HB 09/2016; NM 05/2016). The challenge was not to persuade the Tanzanian government but to make the World Bank stop putting the government under pressure, as Bafadhili elaborated:

We have spoken with the prime minister, vice president. The government knows. But the problem here is the financing agreement between the World Bank and the government. The World Bank always goes to the government and says: 'We agreed that you will [choose] everybody through a tender process. So, I want you to do that. Or else, I will not give you the money for the other projects: for Phase 3, for the flyover in Ubungo, for other projects that you have requested here. We will stop'.

(HB 09/2016)

Hence, UDA wanted to continue with UDA-RT as it has in the last decades; with a privileged role compared to other minibus operators and without the government and international actors like the World Bank interfering (AH 10/2015; HB/NM 09/2016). The re-engagement of the state is particularly decisive for UDA, which had tried to continue to deal with the national government like it always had: indirectly supported by government elites that enabled the company special treatment in the form of exceptional licences and tax exemptions.

Translating a global model is not at all a smooth process; it is hallmarked by tensions and negotiations, detours and standstills. This chapter has revealed the manifold strategies of Dar es Salaam's transport operators, supported by a political elite, to become a permanent part of DART. Tracing the (re)assembling of agreements and their 'detailed details' has delineated how technopolitical controversies have been negotiated in Dar es Salaam. In addition to material and spatial non-compliance, legislative negotiations, the play for time, and a certain degree of self-esteem have restructured power in the DART project. UDA-RT insisted on its exclusive right as a public limited company and refused to operate according to agreements with the state and the World Bank. The forms of resistance that UDA-RT applied – as in-between actors rather than laypersons or experts – go beyond commonly discussed forms of resistance against new technopolitical systems (see Callon et al. 2009; Ureta 2015; Von Schnitzler 2008, Wood 2014, 2015a). Moreover, the specific context is crucial for the emergence of controversies. Local operators refused to collaborate with an international company on the one hand, and with the state on the other hand.

UDA-RT enforced a reformulation of DART's operational model. Through its material practice and physical presence, the company has succeeded in cementing a permanent role within the DART project. Deviations of the global BRT model – as the result of controversies – can also be regarded as productive moments that enforce the order of heterogeneous actors to stay open to negotiation, disagreement and resistance (Law 1992). The process in Dar es Salaam led to significant changes of the presumedly stable plan and to delays in implementation. I will discuss these as 'permanent interim assemblages' in the next chapter. The operational controversies have thus been highly effective: the DART model – BRT in an international PPP structure – has become the UDA-RT model – local elites strategically enforcing their exclusive and permanent rights over operations. That the UDA-RT model would become a best practice for local appropriation of an international project is unlikely for two reasons though. First, the context of the transport sector in Dar es Salaam and the role of UDA is quite unusual and not comparable to other (African) cities. Second, the model lacks global policy mobilisers. However, the hope that the DART model could smoothly circulate to other sites and impact the global BRT model has also been defeated.

Notes

1 *Der Bau ist jetzt nicht so unbedingt spannend. Das ist jetzt kein Nuklearkraftwerk, sondern: man baut das halt. […] Das Spannende, das ist jetzt das zu implementieren.*
2 In 1947, Dar es Salaam Motor Transport Company (DMT), a privately-owned British bus company, was founded. DMT became nationalised in 1970 as part of Nyerere's socialist strategy, and was split into two semi-autonomous transport companies – Shirika la Usafiri Dar es Salaam (UDA) and National Bus Service/*Kampuni ya Mabasi ya Taifa* (NBS/KAMATA, now defunct) – in 1974. The Dar es Salaam City Council (DCC) owned the major shares of UDA, which brought UDA an exclusive licence for bus operations in Dar es Salaam, including subsidies to lower bus fares. UDA's services rapidly deteriorated in the 1970s due to a range of problems (aged and obsolete fleet, lack of qualified personnel and management, etc.), and only revitalised in 2011/2012 after the Simon Group bought the major share of UDA from the DCC (Kanyama et al. 2004; Mfinanga and Madinda 2016; Rizzo 2001; Sohail et al. 2004).
3 The DART Agency only considered the daladala owners but not its drivers and conductors for compensation (see DART Agency 2014c). UDA-RT planned to offer daladala owners to buy UDA-RT shares, which has not been done yet. Hence, both UDA and the Tanzanian government excluded a large part of the previous public transport sector from DART operations. Moreover, most DART drivers used to be lorry drivers before because most daladala drivers who applied were not considered as appropriate DART drivers due to their driving style and their lack of experience with articulated buses. Either they have been located to other routes, or they have continued to drive on the route of Phase 1 because the ISP cannot yet meet the demand. The selection of DART drivers means another deviation from the global BRT narrative – that BRT can integrate drivers of the minibus sector into BRT operations to prevent job losses of the previous public transport system's workforce.
4 BRT systems operate on so-called trunk and feeder routes. Trunk routes connect the BRT terminals and they generally consist of dedicated lanes. Feeder buses serve outside the dedicated lane on a mixed-traffic road.
5 UDA-RT bought 140 buses in total, which is an extra of 56 buses (140 total buses minus 76 buses agreed in the *ISP Agreement*, plus 8 buses as contractual 10 per cent margin of these 76 buses). Two of them arrived earlier in Dar es Salaam as prototypes and for training purposes.
6 *Hawafuati sheria kabisa.*
7 In February 2020, the Tanzanian government once again tendered for a second long-term service provider. When finalising this book (May 2020), no decision has been made (Mirondo 2020).
8 *Walichukua opportunity kama business men, that's normal.*
9 *Bado hatuna uhakika wa mabasi kuanza tarehe 10, kwa sababu bado kuna mambo ambayo hayakamiliki; kama mfumo wa kukusanya nauli, SUMATRA kukamilisha utaratibu wa nauli nk.*
10 *Mwanzo wa Mabadiliko. Fursa kwa wote.*
11 *Sisi ni Watanzania, lazima tutoe huduma.*

References

Akrich, M. 1992. "The De-Scription of Technical Objects". In: W. Bijker and J. Law (eds.), *Shaping Technology/Building Society. Studies in Sociotechnical Change.* Cambridge, MA: MIT Press, 205–224.

Barry, A. 2015. "Infrastructural Times". In: H. Appel, N. Anand and A. Gupta (eds.), *The Infrastructure Toolbox*. Fieldsights, September 24. Accessed at *Cultural Anthropology* (https://culanth.org/fieldsights/series/the-infrastructure-toolbox). Published 24.09.2015, retrieved 17.04.2020.

Barry, A. 2013. "The Translation Zone. Between Actor-Network Theory and International Relations". *Millennium: Journal of International Studies* 41:3, 413–429.

Barry, A. 2012. "Political Situations. Knowledge Controversies in Transnational Governance". *Critical Policy Studies* 6:3, 324–336.

Behrens, R. and S. Ferro 2016. "Barriers to Comprehensive Paratransit Replacement". In: R. Behrens, D. McCormick and D. Mfinanga (eds.), *Paratransit in African Cities. Operations, Regulation and Reform*. Oxon, New York: Routledge, 199–220.

Birch, K. and M. Siemiatycki 2016. "Neoliberalism and the Geographies of Marketization. The Entangling of State and Markets". *Progress in Human Geography* 40:2, 177–198.

Brenner, N., J. Peck and N. Theodore 2010. "Variegated Neoliberalization. Geographies, Modalities, Pathways". *Global Networks* 10:2, 182–222.

Callon, M., P. Lascoumes and Y. Barthe 2009 (2001). *Acting in an Uncertain World. An Essay on Technical Democracy*. Cambridge, MA: MIT Press.

Collier, S. and A. Ong 2005. "Global Assemblages, Anthropological Problems". In: A. Ong and S. Collier (eds.), *Global Assemblages. Technology, Politics, and Ethics as Anthropological Problems*. Malden, MA: Blackwell, 3–21.

DART Agency 2018. "FAQs". Accessed at *DART Agency* (http://dart.go.tz/en/faq/). Published November 2018, retrieved 11.11.2018.

DART Agency 2016. "Invitation to Tender for Procurement of Service Provider for Supply, Operations and Maintenance of Additional buses for the DART System". Accessed at *DART Agency* (http://www.dart.go.tz/). Published 20.06.2016, retrieved 20.07.2016.

DART Agency 2015a. "High Hope for BRT Project as Key Players Agree". Accessed at *DART Agency* (http://dart.go.tz/en/high-hopes-for-brt-project-as-key-players-agree/). Published April 2015, retrieved 20.07.2016.

DART Agency 2015b. "Interim Service Provider Agreement for the Provision of Interim Bus Transport Service for Phase 1 of the DART System, Dar es Salaam, Tanzania". Executive Copy, April 2015. (*unpublished*).

DART Agency 2014a. "Bus Output Specifications". Updated by Transaction Advisory Team. Version 3.2, June 2014. Dar es Salaam. (*unpublished*).

DART Agency 2014b. "Dar Rapid Transit (DART). Project Phase 1. Project Information Memorandum". Final Version, May 2014. Dar es Salaam. (*unpublished*).

DART Agency 2014c. "DART Souvenir on Marketing Consultation Meet on Phase One of the DART Project. Bid to Make a Difference". Accessed at *DART Agency* (http://dart.go.tz/en). Published June 2014, retrieved 20.07.2016.

DCC 2007. "Consultancy Services for the Conceptual Design of a Long Term Integrated Dar es Salaam BRT System and Detailed Design for Initial Corridor. Annex Volume 5. Operational Design". Draft Final Report, April 2007. (*unpublished*).

Farías, I. 2011. "The Politics of Urban Assemblages". *City* 15:3–3, 365–374.

Ferguson, J. 2009. "The Uses of Neoliberalism". *Antipode* 41:1, 166–184.

GoT 2011. "The Public Procurement Act". *Gazette of the United Republic of Tanzania* 52:92, 30.12.2011.

GoT 2010. "The Public Private Partnership Act". *Gazette of the United Republic of Tanzania* 32:91, 06.08.2010.

Hall, D. 2015. "Why Public-Private Partnerships Don't Work. The Many Advantages of the Public Alternative". *Public Services International Research Unit.* University of Greenwich.

Healey, P. 2013. "Circuits of Knowledge and Techniques. The Transnational Flow of Planning Ideas and Practices". *International Journal of Urban and Regional Research* 37:5, 1510–1526.

Hidalgo, D. 2015. "Learning from Delhi's BRT Failure, and Looking to the City's Future". Accessed at *TheCityFix* (http://thecityfix.com/blog/learning-from-delhis-brt-failure-looking-citys-future-dario-hidalgo/). Published 17.09.2015, retrieved 17.04.2020.

Hönke, J. and I. Cuesta-Fernandez 2018. "Mobilising Security and Logistics through an African Port. A Controversies Approach to Infrastructure". *Mobilities* 13:2, 246–260.

Hook, W. 2005. *Institutional and Regulatory Options for Bus Rapid Transit in Developing Countries. Lessons from International Experience.* New York: ITDP.

ITDP 2018a. "BRT Planning 101". Accessed at *ITDP* (https://www.itdp.org/publication/webinar-brt-planning-101/). Published 02.02.2018, retrieved 17.04.2020.

ITDP 2018b. "Dar es Salaam Leads a Breakthrough for African Cities". *Sustainable Transport,* Winter 2018, No. 29. Accessed at *ITDP* (https://www.itdp.org/category/content-type/publication/magazine/). Published 13.02.2018, retrieved 17.04.2020.

ITDP 2017. "The Online BRT Planning Guide". 4th Edition. Accessed at *ITDP* (https://brtguide.itdp.org). Published 16.11.2017, retrieved 17.04.2020.

Jaffe, E. 2012. "Why Are People Rioting Over Bogota's Public Transit System"? Accessed at *CityLab* (https://www.citylab.com/transportation/2012/03/why-are-people-rioting-over-bogotas-public-transit-system/1537/). Published 20.03.2012, retrieved 17.04.2020.

Jasanoff, S. 2012. "Genealogies of STS". *Social Studies of Science* 42:3, 1–7.

JICA 2008. "Dar es Salaam Transport Policy and System Development Master Plan. Final Report". Tokyo. (*unpublished*).

Kanyama, A., A. Carsson-Kanyama, A. L. Lindén and J. Lupala 2004. *Public Transport in Dar es Salaam, Tanzania. Institutional Challenges and Opportunities for a Sustainable Transport System.* Stockholm: Institutionen för Miljöstrategiska Studier.

Kasumuni, L 2018: "Udart Boss: Rethink Revenues Model". Accessed at *The Citizen Tanzania* (http://www.thecitizen.co.tz/News/-Udart-boss--Rethink-revenues-model/1840340-4572088-11bjbpo/index.html). Published 21.05.2018, retrieved 17.04.2020.

Kim, J. 2017. "Laying Foundation Stone for Ubungo Overpass Speech. Remarks by Dr Jim Yong Kim, President of the World Bank Group". Accessed at *Ikulu* (http://blog.ikulu.go.tz/?p=20114). Published 20.03.2017, retrieved 17.04.2020.

Latour, B. 2005. *Reassembling the Social. An Introduction to Actor-Network-Theory.* Oxford: Oxford University Press.

Latour, B. 1991. "Technology is Society Made Durable". In: J. Law (ed.), *A Sociology of Monsters. Essays on Power, Technology, and Domination.* London, New York: Routledge, 103–131.

Latour, B. 1987. *Science in Action. How to Follow Scientists and Engineers through Society.* Cambridge, MA: Harvard University Press.

Law, J. 2004. *After Method. Mess in Social Science Research.* London, New York: Routledge.

Law, J. 1992. "Notes on the Theory of the Actor-Network: Ordering, Strategy, and Heterogeneity". *Systems Practice* 5:4, 379–393.

Loxley, J. 2013. "Are Public-Private Partnerships (PPPs) the Answer to Africa's Infrastructure Needs"? *Review of African Political Economy* 40:137, 485–495.

Malanga, A. 2018. "'We Are Unaware of Udart's 70 Buses at the Port' Says Dart". Accessed at *The Citizen Tanzania* (http://www.thecitizen.co.tz/News/1840340-4576680-2509gxz/index.html). Published 23.05.2018, retrieved 17.04.2020.

Marres, N. 2015. "Why Map Issues? On Controversy Analysis as a Digital Method". *Science, Technology, and Human Values* 40:5, 655–686.

Marres, N. 2012. *Material Participation. Technology, the Environment and Everyday Publics.* Basingstoke: Palgrave Macmillan.

Marres, N. 2007. "The Issues Deserve More Credit. Pragmatist Contributions to the Study of Public Involvement in Controversy". *Social Studies of Science* 37:5, 759–780.

Mawdsley, E. 2017. "Development Geography I. Cooperation, Competition and Convergence between 'North' and 'South'". *Progress in Human Geography* 41:1, 108–117.

McCann, E. 2011. "Urban Policy Mobilities and Global Circuits of Knowledge. Toward a Research Agenda". *Annals of the Association of American Geographers* 101:1, 107–130.

Mfinanga, D. and E. Madinda 2016. "Public Transport and Daladala Service Improvement Prospects in Dar es Salaam". In: R. Behrens, D. McCormick and D. Mfinanga (eds.), *Paratransit in African Cities. Operations, Regulation and Reform.* Oxon, New York: Routledge, 155–173.

Mirondo, R. 2020. "Tanzania Government Seeks New Rapid Bus Operator" Accessed at *The Citizen Tanzania* (https://www.thecitizen.co.tz/news/1840340-5461458-af0wq7/index.html). Published 19.02.2020, retrieved 17.04.2020.

Mitchell, T. 2002. *Rule of Experts. Egypt, Techno-Politics, Modernity.* Berkeley: University of California Press.

Moore, S. 2013. "What's Wrong with Best Practice? Questioning the Typification of New Urbanism". *Urban Studies* 50:11, 2371–2387.

Msikula, A. 2015. "End Not in Sight with DART". Accessed at *The Guardian Tanzania* (http://www.ippmedia.com/frontend/index.php?l=84934). Published 04.10.2015, retrieved 11.11.2018.

Msira, T. 2016. "Why Did Bus Rapid Transit Go Bust in Delhi"? Accessed at *CityLab* (https://www.citylab.com/solutions/2016/12/why-did-bus-rapid-transit-go-bust-in-delhi/510431/). Published 20.12.2016, retrieved 17.04.2020.

Muñoz, J. and D. Molina 2009. "A Multi-Unit Tender Award Process: The Case of Transantiago". *European Journal of Operational Research* 197, 307–311.

Mwangonde, H. 2017. "Make Sure You Make Profit, JPM Tells DART". Accessed at *The Guardian Tanzania* (http://www.ippmedia.com/en/news/make-sure-you-make-profit-jpm-tells-dart). Published 26.01.2017, retrieved 11.11.2018.

Nachilongo, H. 2018. "We'll Get Second Rapid Transit Bus Operator by January, Dart Promises". Accessed at *The Citizen Tanzania* (http://www.thecitizen.co.tz/News/We-ll-get-second-bus-rapid-transit-operator-by-January-/1840340-4757066-wcq3q0z/index.html). Published 13.09.2018, retrieved 17.04.2020.

Nachilongo, H. 2017. "Search for Second Bus Rapid Transit Operator Launched Afresh". Accessed at *The Citizen Tanzania* (http://www.thecitizen.co.tz/News/Search-for-second-BRT-operator-launched-afresh/1840340-3937888-h54hdsz/index.html). Published 23.05.2017, retrieved 17.04.2020.

Nelkin, D. 1979. *Controversy: Politics of Technical Decisions.* Beverly Hills, CA: Sage.

Nelkin, D. 1975. "The Political Impact of Technical Expertise". *Social Studies of Science* 5, 35–54.

Ong, A. 2006. *Neoliberalism as Exception. Mutations in Citizenship and Sovereignty.* Durham, NC, London: Duke University Press.

Paget-Seekings, L. 2015. "Bus Rapid Transit as Neoliberal Contradiction". *Journal of Transport Geography* 48, 115–120.

Parnell, S. and J. Robinson 2012. "(Re)theorizing Cities from the Global South: Looking Beyond Neoliberalism". *Urban Geography* 33:4, 593–617.

Peck, J. and A. Tickell 2002. "Neoliberalizing Space". *Antipode* 34:3, 380–404.

Peter, F. 2016. "Unpaid Taxes Put Brakes on Dar Rapid Buses. Firm Owes 8bn/- in Import Tax". Accessed at *The Guardian Tanzania* (http://www.ippmedia.com/en/unpaid-taxes-put-brakes-dar-rapid-buses). Published 08.01.2016, retrieved 17.04.2020.

Pinch, T. and C. Leuenberger 2006. "Studying Scientific Controversy from the STS Perspective". *Paper presented at the EASTS Conference* 'Science Controversy and Democracy'.

Rizzo, M. and A. Dave. 2018. "Questioning BRTs: A Win-Win Solution to Public Transport Problems in the Cities of Developing Countries"? Accessed at *SOAS* (https://www.soas.ac.uk/registry/scholarships/file126686.pdf). Published 13.02.2018, retrieved 17.04.2020.

Rizzo, M. 2017. *Taken for a Ride. Grounding Neoliberalism, Precarious Labour and Public Transport in an African Metropolis.* Oxford: Oxford University Press.

Rizzo, M. 2014. "The Political Economy of an Urban Megaproject. The Bus Rapid Transport Project in Tanzania". *African Affairs* 114:455, 1–22.

Rizzo, M. 2013. "The Privatisation and Deregulation of Dar es Salaam's Public Transport, 1983–2010. Outcomes and Dilemmas". In: P. Chanie and P. Mihyo (eds.), *Thirty Years of Public Sector Reforms in Africa. Selected Country Experiences.* Kampala: Fountain Publishers, 81–105.

Rizzo, M. 2001. "Being Taken for a Ride. Privatisation of the Dar es Salaam Transport System 1983–1998". *Journal of Modern African Studies* 40:1, 133–157.

Robinson, J. 2015. "'Arriving at' Urban Policies. The Topological Spaces of Urban Policy Mobility". *International Journal of Urban and Regional Research* 39:4, 831–834.

Sauwa, S. and P. Elias 2018. "Dar Transport in Chaos over Udart Troubles". Accessed at *The Citizen Tanzania* (http://www.thecitizen.co.tz/News/Dar-transport-in-chaos-over-Udart-troubles/1840340-4800784-59ig8sz/index.html). Published 11.10.2018, retrieved 17.04.2020.

Schalekamp, H., A. Golub and R. Behrens 2016. "Approaches to Paratransit Reform". In: R. Behrens, D. McCormick and D. Mfinanga (eds.), *Paratransit in African Cities. Operations, Regulation and Reform.* Oxon, New York: Routledge, 100–125.

Siemiatycki, M. 2013. "The Global Production of Transportation Public-Private Partnerships". *International Journal of Urban and Regional Research* 37:4, 1254–1272.

Siemiatycki, M. 2011. "Urban Transportation Public-Private Partnerships. Drivers of Uneven Development"? *Environment and Planning A* 43, 1707–1722.

Sohail, M., D. Maunder and D. Miles 2004. "Managing Public Transport in Developing Countries. Stakeholder Perspectives in Dar es Salaam and Faisalabad". *International Journal of Transport Management* 2, 149–160.

Stengers, I. 2005. "The Cosmopolitical Proposal". In: B. Latour and P. Weibel (eds.), *Making Things Public. Atmospheres of Democracy.* Cambridge, MA: MIT Press, 994–1003.

Temenos, C. and T. Baker 2015. "Enriching Urban Policy Mobilities Research". *International Journal of Urban and Regional Research*, 841–843.

The Citizen Reporter 2017. "High Court Halts Search for Second Dart Bus Operator". Accessed at *The Citizen Tanzania* (http://www.thecitizen.co.tz/News/High-Court-halts-search-for-second-Dart-bus-operator/1840340-3789404-tyta1nz/index.html). Published 27.01.2017, retrieved 17.04.2020.

Ureta, S. 2015. *Assembling Policy. Transantiago, Human Devices, and the Dream of a World-Class Society.* Cambridge, MA: MIT Press.

Ureta, S. 2014. "The Shelter that Wasn't There. On the Politics of Co-Ordinating Multiple Urban Assemblages in Santiago, Chile". *Urban Studies* 52:2, 231–246.

Valeriani, S. 2010. "Facts and Building Artefacts: What Travels in Material Objects"? In: P. Howlett and M. Morgan (eds.), *How Well Do Facts Travel? The Dissemination of Reliable Knowledge.* Cambridge: Cambridge University Press, 43–71.

Venturini, T. 2009. "Diving in Magma: How to Explore Controversies with Actor-Network Theory". *Public Understanding of Science* 19:3, 258–273.

Vigar, G. 2017. "The Four Knowledges of Transport Planning: Enacting a More Communicative, Trans-Disciplinary Policy and Decision-Making". *Transport Policy* 58, 39–45.

Von Schnitzler, A. 2008. "Citizenship Prepaid: Water, Calculability, and Techno-Politics in South Africa". *Journal of Southern African Studies* 34:4, 899–917.

Whatmore, S. 2009. "Mapping Knowledge Controversies. Science, Democracy and the redistribution of Expertise". *Progress in Human Geography* 33:5, 587–598.

Wood, A. 2015a. "Competing for Knowledge. Leaders ad Laggards of Bus Rapid Transit in South Africa". *Urban Forum* 26, 203–221.

Wood, A. 2015b. "Multiple Temporalities of Policy Circulation. Gradual, Repetitive and Delayed Processes of BRT Adoption in South African Cities". *International Journal of Urban and Regional Research* 39:3, 1–13.

Wood, A. 2014. "Moving Policy. Global and Local Characters Circulating Bus Rapid Transit Through South African Cities". *Urban Geography* 35:8, 1238–1254.

World Bank 2017. "#Blog4Dev: How Can the Private Sector Help Deliver Better Public Services in Tanzania"? Accessed at *The World Bank* (http://www.worldbank.org/en/events/2017/12/12/blog4dev-how-can-the-private-sector-help-deliver-better-public-services-in-tanzania). Published 12.12.2017, retrieved 17.04.2020.

World Bank 2016. "Tanzania Economic Update: More, Better Infrastructure, Increased Human Investment and Renewed Public-Private Partnerships Key to Poverty Reduction in Tanzania". Accessed at *The World Bank* (http://www.worldbank.org/en/country/tanzania/publication/tanzania-economic-update-the-road-less-traveled-unleashing-public-private-partnerships-in-tanzania). Published May 2016, retrieved 17.04.2020.

6 Permanent interim assemblages

When ITDP presents the global BRT model to its audience, it likes to show a table titled 'Bus Operations Best Practices', which represents the choices of several cities in terms of fare collection and the inclusion of existing operators. The table indicates whether or not the bus operator was chosen through a competitive tender. All BRT systems give a clear answer in each category: Yes (green) or No (red). Preparing for the Sustainable Urban Transport Conference in Nairobi, the ITDP director of the Africa Office wants to include DART as best practice in this table. He marks the field for 'competitive tender' in green: 'Yes'. I insist: DART operations have not yet been tendered, and it is not certain that they will be tendered – even if this is what the plans and contracts say. The director agrees and marks the field red. Because he cannot bring himself to replace the 'Yes' with 'No', DART is now the only BRT system on the table that does not give a clear answer: 'No/Yes' is written in the red field. Is it somewhere in between? Still to be decided?

(November 2016)

Controversies materialise and affect DART's operations. As shown in the previous chapter, UDA-RT's (Usafiri Dar es Salaam Rapid Transit) non-compliance with agreements has led to ongoing changes and multiple temporalities of DART. In order to endure these constant changes, DART's shape has had to be flexible. The operational model has proved to be mutable, gradually changing from permanent to interim and back again. Besides this major transformation of the operational model, a myriad of other changes shows similar tendencies: things have become permanent that were planned to be interim. Based on that, this chapter discusses the intermittent assembling of DART's operations by following the arguments that 'temporalities of assemblages are emergent' (Collier and Ong 2005: 12), that infrastructure networks are 'precarious achievements' (Graham and Marvin 2001: 182), and that infrastructures have in parallel 'constructive and destructive moments' (Howe et al. 2015: 6). By emphasising the constant (de)stabilising and (re)making of the DART project and how this evolves in practice, the mutual shaping of planning and implementation comes to the fore. Technopolitical assemblages are flexible rather than static; they generate new meanings and effects, and always embody a certain degree of unpredictability (Hecht 2011: 3, 11).

The assembling of DART's central components was far from a continuous process; it happened at different tempos, unexpectedly pausing at various points of time. Some components have stabilised while others continue to be constantly debated, negotiated and modified. The processuality itself becomes apparent not only in DART's overall delays, but also in particular moments of deferral, repetition and standstill; moments that are ahead of (the planned) time, and moments that were not considered at all. Hence, planning is a messy procedure because 'the reality is different than the plan' (LL 11/2016). Also, mobilising policies is not as stringent as it might seem. Even though 'policy time is speeding up' (Peck and Theodore 2010: 172), global policies do not circulate smoothly, and assembling those policies in a specific locale is a slow and hesitant process. The delays in Dar es Salaam contradict one of the main arguments for BRT, i.e. a rapid and low-cost implementation. Thus, Wood (2015: 11) criticises the narrative of the linear and straightforward realisation of BRT projects and demands that scholars give attention to slow policies instead of fast ones.

The multiple temporalities of the DART process are reflected in its narrative. The DART Agency captioned a handout for a market consultation meeting with 'DART: The Long March' (DART Agency 2014c), and ITDP ran an article in its annual magazine titled '14 Years in the Making, DART Brings Mobility to Dar es Salaam' (ITDP 2017a). Steps did not just slow down; some suddenly went faster than anticipated, overtaking other steps within the process and changing the chronology. Therefore, I conceptualise the temporalities of DART's assembling as being permanently interim, and I shift the focus from fast versus slow policies to the heterogeneous process of assembling travelling models. Based on the most remarkable examples drawn from observing DART and participating in the system's usage, this chapter illustrates the multiple, gradual, twisted, repetitive and delayed temporalities that make DART permanently interim. Focusing on the new spatialities and temporalities in Dar es Salaam, I describe the gradual implementation and (re)making of the system and illustrate the ad-hoc and reactive manner of disciplining and stabilising DART. Furthermore, I question the clear differentiation between daladala and DART, because the two systems have much in common when looking closely at DART's everyday practices.

(De)Stabilising and (re)making

Infrastructure assemblages are heterogeneous and unstable; they flow through space, occupying a 'continuous unreliability' (Graham and

Thrift 2007: 10). Their temporality makes them ephemeral, permanently (de)stabilised (Bowker 2015; Larkin 2008). DART itself is heterogeneous, contingent, unstable, partial and situated. It is not a ready-made product, and the relationships between the elements are characterised by inherent tensions and critical reflections (Collier 2006: 400). An assemblage like DART will thus never be completely stable; it exists in a constant (re)making over time and space (Callon and Law 2004: 3).

Scripts

Scripts – the products of an innovator's vision about 'the world in the technical content of a new object' (Akrich 1992: 208) – of a policy assemblage support the assemblage's standardisation and stabilisation (Ureta 2015). Inscription is linked to displacement because it is the final stage of a mobilisation process (Latour 1986). Inscriptions aim to increase both the mobility and immutability of things because they 'allow new translations and articulations while keeping some types of relations intact' (Latour 1999: 306). Moreover, scripts assign different but interrelated roles to the (non)human actors of an assemblage. When engineers design a technology, they try to anticipate the environment of a technology and user behaviours by inscribing the roles of (technical) objects in their material composition. Inscription is both a technical and a social process (Akrich 1992; Von Schnitzler 2008: 912).

Users of DART are heterogeneous: they comprise not only passengers but also drivers, station attendants, and operation managers, among others. None of these users necessarily comply with the pre-defined scripts. Sometimes, they employ objects in a different manner and 'de-scribe' the role of technical objects:

> Even in the case of quite sophisticated and powerful scripts there is always space for unexpected outcomes, whether as a result of conscious resistance by the individuals or organizations being affected or simply because of unexpected nonhuman agencies.
>
> (Ureta 2015: 9)

As the opposite of inscription, de-scription is the 'mechanism of adjustment between the imagined and the real user' (Akrich 1992: 209; see also Akrich and Latour 1992: 259). De-scription necessarily leads to a deviation from, and a reconfiguration of, the script over time, i.e. between initial design and actual employment. Whereas some scripts of BRT are stable and cannot easily be altered (e.g. the standardised platform height of a BRT station), other scripts change throughout the process of design, operation and maintenance (Pineda 2010, 2011). As roles perpetually change, being inscribed and de-scribed, scripts are present and absent at different points of DART's planning and implementation process. De-scriptions, leading to mutation, are more the norm than the exception in this process.

Every actor of an assemblage has the power to resist their inscribed role and de-scribe them according to their abilities and desires. In terms of their initial and prime function (see Latour 2005: 70–71), buses transport people, turnstiles control the validity of tickets, and stations serve as an interface between passengers and buses. However, even though these functions had not been in place in Dar es Salaam for long, they have repeatedly been questioned. Diverse actors have been shaping the complex process of DART's implementation: not only political decisions, but also everyday practices and nonhuman agency have reconfigured DART's script. The most striking example is UDA-RT's resistance to the power and liability of contracts described in the previous chapter, which the transaction advisor had previously inscribed into these documents. Furthermore, DART's environment and users' behaviours changed as soon as DART came into being. Planners and manufacturers defined the role of buses to transport passengers, but DART's buses took on various roles at different points of time and space (Jacobsen 2016); their assembling has involved Tanzanian traffic law, BRT standards and specifications of DART's infrastructure, as well as expertise from the Chinese manufacturer and South African and German gearbox specialists, and technologies from various places (EM 09/2015; LH 10/2015). Furthermore, the buses might appear as models in policy papers of international consulting firms, in technical instructions for the bus manufacturer, as materialised prototypes arriving in Dar es Salaam, as a bus fleet being expressed in the controversial number 140, as modern means of public transport, as sky-blue vehicle bodies with traces of usage and accidents, as fast and huge vehicles impressing passengers while making minibuses look slow and small, or as the types articulated bus and rigid bus.[1]

The DART Agency (2014a) defined the buses' technical shape in the *Bus Output Specifications*, aiming to inscribe the roles of the buses and their interactions with users. These roles encompass values about how to offer comfort, safety and reliability. However, drivers, light barriers, passengers, speed limits and the road surface do not comply entirely with their inscribed roles. For instance, drivers ignore the warning signal that appears when they speed, doors risk injuring passengers when they open and close, and passengers do not sit and stand in the buses how and where they should. Comfort is drastically reduced due to overcrowding and the lack of air conditioning. But DART's buses are adaptable and flexible; they transport passengers above the maximum capacity, and feeder buses allow (dis)embarking at elevated stations on the left and at the kerbside at the right. The buses and other elements of DART's physical infrastructure are reminiscent of the fluidity of the bush pump and other multiple objects (De Laet and Mol 2000).

The linear space of 21 kilometres of ferroconcrete has changed the shape of the city, crossing the city from west to east. As other works have shown, concrete of large transport projects can change the relation between state and civil society (Harris 2013; Harvey 2010; Harvey and Knox 2015). Moreover, these projects and materials can raise questions of speed, durability and modernity. In Dar es Salaam, people had quickly integrated themselves into DART's physical infrastructure and de-scribed it according to their needs, even before operations had started:

> Bus stations have become shelters for homeless and drug addicts, the BRT corridor a motorbike highway on Morogoro Road and a huge market area at lower Msimbazi Street. Every day, vendors turn crash barriers into stalls to present their goods like roasted cassava, batteries and underwear. These pillars also serve for election campaign posters (see Figure 6.1). The feeder stations at Morocco have become a training ground for children learning to ride a bicycle, and the upper Msimbazi Street is used as parking area for private cars and daladala.
>
> (October 2015)

Figure 6.1 Additional functions of a crash barrier: billboard and pedestal for a grill (author's photo 10/2015).

After the start of operations, DART's infrastructure has continued to be used in multiple ways, thus de-scribing its initially inscribed roles: materials were present and absent in various ways, enabling and disabling diverse forms of usage.

Present absences

Presence and absence are interdependent and not opposed to one another (Callon and Law 2004; Meyer 2012). Thus, scripts can be present in one shape and absent in another. A script of an object can be present in its (physical) absence (e.g. no bus operation without registered buses, materialised through a number plate), or become present through the presence of another object (e.g. the logo of the DART Agency, which had been absent on the body of the bus but present in the *Bus Output Specifications*). The presence or absence of an object can modify the script of an assemblage because both – presence and absence – have their own agency and performativity (Hetherington 2004; Meyer 2012). If scripts are not followed in such a way that infrastructures break down, the presence of the latter is felt in their absence (e.g. when water does not flow from the tap or electricity does not come from the socket, or when there are no buses running on the streets). The presence of absence is experienced most acutely in practices and emotions (Frers 2013: 434), and a breakdown leads to 'more detailed understanding of the relational nature of infrastructure' (Star 1999: 382). Thus, unfunctional infrastructures become more present for the users. Even infrastructures that have never been materially present can also be felt in their absence, existing in imaginations and discourses. For instance, many passengers only experienced the absence of safety features on daladala after they had experienced DART's safety politics.

DART's assembling has shown that the experience of a present absence changes practice. The transport system has been present in various, temporally changing forms during its planning and implementation process. Also, DART's buses appeared in different forms of presence and absence. Even before they were manufactured, the sky-blue buses were present in people's imaginations, in public discourse and the media, as well as in documents like the *Bus Output Specifications* and the *ISP* (interim service provider) *Agreement* (DART Agency 2014a, 2015). When the construction of the physical infrastructure had been finalised and the *ISP Agreement* was signed, the buses, bound up in a legal dispute over customs duties, were actively missing. Their physical presence went through different stages, so that after touching the BRT corridor in September 2015 for their initial ride from Dar es Salaam port to the terminals in Kivukoni and Gerezani, the buses disappeared again – apart from the two prototype buses that were used for training drivers (see Chapter 5).

I realised the unresolved controversy between the national government, the World Bank and UDA through the absence of number plates. I saw 138 out of 140 buses without number plates parked at the terminals before I officially came to know about the clearance issues between UDA-RT and the Tanzania Revenue Authority (TRA). Number plates are the materialisation of formal registration and tax payment; they give a vehicle a legal status, proving that it was imported legally and that import duties were paid. Only with that legal status were DART buses allowed to operate in public space. A number plate is hence not technically but legally necessary for the bus to fulfil its role as a means of public transport. Since the number plate is an indispensable component of the DART script, its role is not flexible. Unlike all the other material components of the bus, the number plate cannot be replaced with spare parts from elsewhere. The material inscription of laws and state power is crucial for the functioning of the DART script; the state is materialised in the number plate so that the number plate directly connects the vehicle with the state. This means that the temporarily missing number plates stand not only for missing tax payments and legal registration, but further show the absence of the Tanzanian government's support in respect of UDA-RT's attempt to strengthen its position. Hence, negotiating (state) power was done via DART's material components. The physical absence of buses on the corridor and the number plates on the buses became a political issue. Until April 2016, when they were released from the TRA to undergo test drives, the buses did not move, but nevertheless they were present in political negotiations and public discourse; everyone was waiting for them to be registered so that operations might finally begin. Therefore, absences and presences, both in material and discursive shapes, provide insights into technopolitical dimensions of planning, implementing and stabilising infrastructures.

The immobility of the DART buses without number plates symbolises the standstill of the planning process. Nevertheless, the process has never been at a complete standstill. Likewise, 'stillness is thoroughly incorporated into the practices of moving' (Cresswell 2012: 648), and moments of standstill in time (regarding the process) and in space (regarding physical movements) complement each other. At times in between obvious movements, things that might seem absent from the outside have only moved to the background. DART is constantly mutating in time and space – not only at times of open progress expressed in milestones reached and in ceremonial events. Moreover, the multitude of things that have not come into being as planned illustrate the tensions between present absences and parallel temporalities. In 2018, stakeholders attempted to subsequently add components to the physical infrastructure that seemed to be missing. A wave of retrofitting swept over the system: ITDP mounted the new station signage (directly prior to the *Mobilize* summit), UDA-RT demanded that the DART Agency install more sanitary facilities at terminals and stations, and the municipality developed parking facilities at Kimara terminal. Other parts were removed from DART's script during the planning process. Most importantly, the stations of Phase 1 will most likely never be equipped with real-time displays or

automatically sliding doors, as initially planned. In 2015, several stakeholders of DART still believed that these automated doors would materialise.
Others had stopped dreaming about them because electrifying and maintaining those six hundred doors including a guaranteed energy supply and
energy storage would be too costly (DS/MH 09/2015; HM 10/2015). Today,
station attendants continue to manually open the doors in the morning and
close them in the evening.

Standstills and shifts

> 'Welcome to udart, Fast reliable and cheap': On the start page of
> UDA-RT's website, visitors can insert their e-mail addresses to be
> informed once operations start. But the function does not work be-
> cause the page shows a countdown that has not yet started. Instead,
> it is already all on zero: 00 days, 00 hours, 00 minutes, 00 seconds –
> as if UDA-RT is starting its services the second you enter the home-
> page. A few months later, the homepage is offline again.
>
> (March 2016)

Viewed in hindsight, DART moved slowly but surely. However, at certain
times in the process, uncertainty became serious. And DART has not yet
been completed, and it never will be; DART is continuously in the making. When did DART become 'real' (Latour 1996: 85), when did the system
come into being? With its construction, the start of the interim service operations, the official inauguration by the Tanzanian President – or will it
only be when Phase 1 becomes fully operational with two bus operators
and a fleet of 305 buses? Different points in time appear as milestones of
DART's emergence, most importantly the multiple inaugurations of Phase 1
and the different standpoints concerning the inception of the ISP operations
in times of political pressure. In 2016, arguments for a small start (UDA-RT
pays import duties in instalments so that buses can be phased in gradually)
contrasted with arguments for a big start (the ISP only begins operating
after the 140 buses and the AFCS (automated fare collection system) are
fully cleared and installed). Government officials initially supported the big
start, assuming that risks would be reduced to a minimum. As the DART
Agency's director of operations and infrastructure management elaborated:
'If we start poorly (*tukianza vibaya*), people will throw stones on us' (RL
10/2015), and an international BRT consultant declared: 'It's a problem if you
don't begin well. You need to get a good first line' (LL 11/2016). UDA-RT,
too, was in favour of a big start, deciding against gradual payment for the
release of the buses and the AFCS equipment, even though this would have
enabled it to generate revenue earlier.

After another round of negotiations and delays, DART's stakeholders
changed their mind. The World Bank advised against a 'big bang' since
UDA-RT already had 'ten balls in two hands', as James Kombe put it (JK

05/2016). Also, the DART Agency became convinced that the idea of a perfectly working DART system was an illusion. In addition, pressure had been growing on UDA-RT to generate revenue in order to repay loans, and on the DART Agency to finally satisfy numerous expectations raised by Dar es Salaam's commuters, the Magufuli government, the World Bank, ITDP and others. The small start became the new goal. According to the director of operations and infrastructure management of the DART Agency:

> They need to start! It is not appropriate to stand still. (*Haipaswi kusimama.*) They don't need to be perfect. Start with what we have. [...] If we waited for perfection, we couldn't have launched before June or August. [...] They have to start somewhere. And then, they work on the efficiencies.
>
> (RL 05/2016)

Thus, UDA-RT started operations without having either a fully installed AFCS or a control centre in place. For the CEO of the DART Agency, the key lesson for cities interested in BRT was to set up an operational design early, and to have the operators ready to begin before the construction works are completed, so that no one can put the government under pressure and renegotiate conditions (GV 11/2016).

Hence, much back and forth characterised the projected inception of operations, as well as the official inaugurations of Phase 1. Beginning operations involved two, partially overlapping, aspects: the question of the operational start by the ISP on the one hand, and the inauguration of the full Phase 1 on the other. Due to the controversial times that involved restructuring the operational model in 2015 (see Chapter 5), forecasts about the actual inception of UDA-RT's bus operations changed from year to year, month to month. Then, in September 2015, just before the upcoming presidential elections, the arrival of the buses at the port seemed to push the process of DART's emergence forward. A transport engineer from the DART Agency told me how this directly affected the discussions on launching DART: 'Because the buses are here, we need to start' (RL 10/2015). In contrast, an external transport consultant reflected on the unrealistic plan to start before presidential elections:

> The first thing is the completion of the basic infrastructure. The stations are not complete. There are still some road works going on. Signalling system is not in progress. The earlier plan was to start sometime in the third week of October, and we have just left three weeks' time. And during the World Bank mission, I have also come to know that there is a lot of problems at the depots, signalling system is not working, the wiring has not been, controllers are not there. So, I don't think, in two, three weeks they will be able to do it. And without a signalling system, proper signal system, no BRT will work because there are a number of

junctions where manual stopping, manual control of the traffic is not possible.

(AM 09/2015)

Before October 2015, the elections drove the ruling party to launch Phase 1 – particularly because the government justified its decision to employ an interim service with the argument that it would speed up the start of operations (HM 03/2015; Mwangonde 2015). But after controversies had already arisen, the elections themselves became a hindrance to launching Phase 1, because too many issues still needed to be solved. The government did not want to risk a start of operations that might involve unpredictable issues, which could undermine the project's – and hence the government's – reputation. The difficult situation created by the purchase of many buses and the AFCS made opposing stakeholders feel as if they depended on steps taken by the other side. Whereas the national government waited for the elections to come and for UDA-RT to explain their large purchase, UDA-RT officials hid from the public and the media in order to avoid doing just that. Nevertheless, UDA-RT accused the government of being responsible for this standstill. According to UDA-RT's spokesman, the company was 'being caught between a rock and a hard place' (AH 10/2015), since the government had not released the buses and thereby prevented the targeted start of the ISP operations. This standstill also affected the driving instructors, the (future) passengers and drivers, as well as the team of the Chinese bus manufacturer that was based in Dar es Salaam for after sales services (DH 10/2015; SC 05/2016). Politicians were not permitted to take financial decisions right before elections, and DART was put on hold. After the elections, DART became the showcase project for the anti-corruption politics of the new President John Magufuli (see Chapter 4).

Inaugurations are crucial for infrastructure politics, and the Tanzanian government inaugurated DART more than once. During these public events, high-ranking officials from the government and the World Bank scanned paper tickets and smart cards, walked through turnstiles, had a ride on a bus or even drove the bus themselves (DART Agency 2017; Mwisaleo 2015). Besides the official construction launch by the former Tanzanian President Jakaya Kikwete in September 2010 and the welcome event for the two prototype buses with high-ranking government officials in August 2015, two major inauguration events took place. The first was held almost punctually on the second day of operations. Accompanied by numerous journalists and bystanders, the regional commissioner strode through a turnstile at Gerezani Terminal (see Figure 6.2) and took a ride in one of the buses. In his speech, he honoured DART, made certain rules clear and emphasised the importance of DART for the city and its residents: infrastructure politics at work.

The second, main inauguration, executed by the President, was postponed to January 2017, eight months after the start of the ISP operations,

Figure 6.2 Dar es Salaam's regional commissioner scans a ticket, assisted by the CEO of UDA-RT (author's photo 05/2016).

undermining the idea that an inauguration initiates, and precedes, something new. The reasons for this delay remain vague. The CEO of the DART Agency argued that it was better to have this representative event once everything will be fully installed:

> We would prefer if he [Magufuli] came at a time that we already have smooth operations, when we have already solved most of the challenges. [...] This inauguration will happen, but when exactly? The inauguration doesn't mean it has to be before start of operations. We will do it while the operations are going well.[2]

(GV 05/2016)

There is reason to believe that the World Bank influenced the date of the event. The bank only wanted to create international attention once operations had become relatively stable and Phase 1 looked like a successful investment.

The inauguration itself was ostentatious. DART operations definitely made a better impression than during the inaugural event in 2016. Images and videos of Magufuli making a speech, striding through a turnstile at Gerezani Terminal, driving a DART bus, hooting to the crowd, and chatting with World Bank officials went viral on several multimedia sites and were also printed on the cover of the UDA magazine (UDA 2017). A sign, in Kiswahili and English, was installed at Ubungo:

> Phase 1 of the Dar es Salaam Bus Rapid Transit (BRT) infrastructure and bus operations have been officially inaugurated by his Excellency Dr. John Pombe Joseph Magufuli, President of the United Republic of Tanzania on 25th January 2017 in the presence of Mr. Makhtar Diop, World Bank Vice President, Africa Region.

Several strategic steps ultimately followed the inauguration in January 2017: the World Bank secured funding for DART's Phases 3 and 4; the Simon Group finally paid its debt to the national government for its 51 per cent holding in UDA that it had bought in 2011; and UDA-RT brought the case concerning the tendering of a second service provider to court (Masare 2017; World Bank 2017). In conclusion, these multiple inaugurations of DART show that the process of planning and implementing a transport system does not follow a linear order. Timing is largely the result of political agendas.

Mess or success?

On several occasions, the success of DART has been questioned. Referring to DART's delays, the chairman of UN Habitat contradicted the project's best-practice narrative at the transport conference in November 2016: 'What happened in Tanzania – 14 years – is quite shocking. […] It is also difficult for investors to work with this unpredictable time frame'. Global consultants started to speak not only about BRT as a solution to the mess (BN 03/2015; LL 11/2016), but also about DART as a mess itself. However, only Herbert Meyer, the former chief technical advisor of the DART Agency, used the term 'failure (of the world-class system)' (HM 05/2016). Other global consultants criticised the 'doing cheap' of UDA-RT's operational service, which stood in contrast to its very good physical design (JB 09/2016; LL 11/2016).

During the first operational week, when rides were for free to attract passengers and create public acceptance for DART, people intensively used the service. Some were even just riding up and down the corridor for fun, without a specific destination. Consequently, the then CEO of UDA-RT and the regional commissioner had to rebuke the excited passengers. UDA-RT established the rule that drivers had to completely unload the bus at the terminal before letting new passengers embark on the bus. However, the international recipe had perfectly worked out: people accept and use the new BRT system. After the introduction of fares, overcrowding persisted – at least during peak hours – and has been widely criticised. The journalist Mngodo (2016) described the purchase of a ticket as a 'nightmare'; passengers had to fight their way through to the ticket booths or be patient in long queues.

I finally catch an overcrowded bus at Jangwani directing Ubungo after that two buses had passed without stopping at the station. It is peak hour. Experienced passengers give advices to rather new passengers: 'Do not stand in the yellow area! The door will hurt you!' (*Mlango utakuumwa!*). The light sensors are off so that doors open and close even if people block the *No Standing Area*. A few stops later, the bus driver calls the passengers to squeeze a little more because at the centre of the bus is a bit of space left. Then, abruptly, the driver stands on the brakes, probably someone un-predictably invaded or crossed the corridor. The braking pushes the passengers to the front. A passenger jokes: 'The driver tried to squeeze people to the front.' (*Alitaka kusogeza watu mbele.*) People laugh.

(September 2016)

Clearly, DART operations do not meet the image of a high-quality, comfort-able and reliable public transport system. Overcrowding is another expres-sion of the mutating DART script that contradicts the global BRT narrative (Giliard 2017). Buses lack safety equipment like window hammers and fire extinguishers so that when buses caught fire in December 2018 and Septem-ber 2019, passengers found themselves in dangerous situations (Namkwahe 2019). Moreover, DART's services have a fundamental problem with delays and infrequent bus departures, and also the comfort inside a DART bus reminds at daladala: either people get a ticket and a few square centimetres in a sticky hot bus – sometimes after waiting up to one hour at a station until one of the passing buses actually has some space for additional passengers – or they give up and switch back to other modes of transport. DART's over-crowding is caused not only by continuously increasing passenger numbers and a lack of substantial scheduling, but also by the ISP's reduced bus fleet. The DART Agency initially announced that it would financially penalise UDA-RT for each bus that exceeds the maximum passengers load of 105 per cent (AM 09/2015), which was impossible to implement because most buses exceeded the maximum capacity. UDA-RT benefits financially from bus shortages and the resultant overcrowding: the higher the passenger loads, the higher the revenue for the bus operator; or in the words of global BRT consultant Luc Lavoie: 'low price, low quality' (LL 11/2016). DART is not the only BRT system that experienced difficulties with overcrowding. Years after implementation, the prestigious *Transmilenio* is also continuously overcrowded, which Lavoie interprets as that *Transmilenio* has become a 'victim of its own success' (LL 11/2016). From that angle, the matter of over-crowding is Janus-faced. On the one hand, overcrowding stands for the suc-cess of a BRT system (i.e. public acceptance of the project and revenue for the operators). On the other hand, it stands for operational difficulties.

DART's multiply emerging forms are also expressed in the premature ageing of the dedicated lane, buses and stations. On a late evening in March

2016, two months prior to start of operations, a long-distance bus on the way to the bus terminal in Ubungo accidentally drove into the DART station Baruti. This picture went viral, showing the bus stuck within the passenger waiting area of the station as if it was a garage. Stories suggest that the driver was either drunk or tired after a long working day. Whatever the case, how could the bus enter so easily the DART corridor that was normally protected by concrete barriers to prevent vehicles invading DART's infrastructure? While this question remained unanswered, the process of repairing bus and station has revealed a prolonged process characterised by negotiation and change. The Tanzanian government, as the owner of the station responsible for repair and maintenance, obtained tenders for its repair, which the bus company had to pay (MH 09/2016; RP 05/2016). But because the bus company found the offers too expensive, it repaired the station itself. A road engineer who had been responsible for the initial constructing of the station described the result as follows:

> Fair enough, I admit: If you pass [the station] and you don't look at it as an engineer, you only see that something is wrong. But the details, you only see them as an engineer. [...] In Germany, this would not be accepted because of the damaged structure. This means: the whole station has moved to the right, and some day they will ask if the statics will withstand.[3]
>
> (MH 09/2016)

During the 18-month period of negotiations and reconstruction works, the station could not be used. However, UDA-RT drivers flexibly handled the situation until DART's leadership clarified:

> UDA-RT, you are asked to inform your drivers [...] not to service the station BARUTI, it is broken. They should stop dropping off passengers since it is dangerous and threatens the passengers' safety. They drop them off outside the station at raised platform, then passengers jump down on the road in a risky area for them.[4]

Other similar incidents reveal material components deteriorating, being destroyed, degraded or decayed even before and shortly after the ISP had become operational. The gravest story concerns the regular inundation of Jangwani Valley, located between the city centre and Magomeni district and accommodating the DART headquarters, the bus depot and workshops, and a part of the BRT corridor including a bridge that crosses Msimbazi River. Twice a year, rains damage workshops and buses and cause disruption of the services, both leading to massive financial losses. This issue is also characterised by unexpected outcomes that question the initial plans and lead to moments of deferred temporalities. Even though heavy rains in March/April and October/November are seasonal and thus predictable in these latitudes, rain was not sufficiently calculated in DART's infrastructural design. During inundation, when operations come to a standstill for up to several days, operational schedules are even more absent than usual. In November 2017, heavy rains caused particularly tremendous damage, leading to a financial loss for UDA-RT due to lost revenue and damage to more than a third of its fleet (Nachilongo 2017). The local government recommended using daladala, which were able to take another route to the city centre because they are not bound to the corridor that passes the inundated Jangwani Bridge. In April 2018, the Tanzanian media coverage of DART became increasingly negative. Newspapers ran headlines such as 'Tanzania admits BRT project blunder' (Mbwiga 2018) and 'If you're in Dar es Salaam, don't try to travel to these places today!' (listing Jangwani at first place and displaying a photo with DART buses stuck in, and people wading through, the dirty water) (The Citizen Reporter 2018a, 2018b). These articles express the counter-concept to mobility, velocity and reliability – key words in the global BRT narrative, reflecting the ideals and expectations on DART.

Why is Jangwani so vulnerable? The bus depot was located there because it was the only open space along the corridor that could accommodate several hundred buses (Ka'bange et al. 2014: 182). Inundation is nothing new in the valley, but various actors had not expected such extreme flooding. Msimbazi River is highly polluted, which increases the severity of flooding. However, primarily due to cost constraints and insufficient data, the initial design did not contain provisions to raise either the bridge or the depot. The road agency TANROADS argued that the distance from the riverbed to the bridge was more than four metres in the dry season, which could only become a problem during heavy rains (PF 10/2015; WG 10/2015). Even though the construction company warned the Tanzanian government and suggested amending the design, TANROADS considered the occasional flooding of the bridge and the depot to be tolerable and did not approve any such changes:

> The principle is: the entrepreneur builds according to the plan that the engineer has approved for construction. [...] The existing bridge should only be widened in any case. Technically correct would have been to demolish the bridge and reconstruct it higher. Which of course would have led to a cost explosion again. And because the design of the bus depot had been adjusted to the height of the existing bridge, they said: 'If not this, then also not that, and that.'[5]

(MH 09/2015)

Construction companies widened the bridge and built the depot according to the initial design (DS/MH 09/2016; RO 03/2015). Due to the manifold problems emerging from this alleged cost-saving measure, Jangwani became highly political and stakeholders suffered repeated accusations. The government ordered various measures: cleaning the river, digging a ditch and building a concrete wall around the depot. These small solutions reduced but did not solve the problem. As the chief engineer of the contractor told me: 'The problem will be: The buses, although they will be safe within the compound, they will not be able to get out of the compound onto the road' (WG 10/2015). Global consultants continued to search for a long-term solution to this 'sensitive topic', as they described Jangwani. At the time of writing, four years after inaugurating the system, no permanent solution has been found, and people seem to have grown used to seasonal interruptions of DART operations (Jumanne 2019; The Citizen Reporter 2020).

In order to hold off the 'constant decay of the world' (Graham and Thrift 2007: 1–2), infrastructures require timely and continuous repair and maintenance, a 'humble but vital process'. Barry (2015), referring to Jackson (2014, 2015), argues that infrastructures demand regular monitoring and repair. Because they break, corrode and are sabotaged, they 'are not the stable base on which a political superstructure can be established'. In Dar es Salaam, the interim use of the stations by other actors and the forces of nature had left traces that had to be eliminated before the inception of operations. Stations were full of litter and the corridor was full of little hills of sand and dust. After one month of operations, Magufuli complained about the damage to the physical infrastructure, which cost the Tanzanian taxpayers billions of Tanzanian Shillings (ITV Tanzania 2016). The need to maintain and repair was not sufficiently thought through, as the CEO of the DART Agency admitted:

> We put in the infrastructure but we forget about maintenance. And in the maintenance, we need to plan it well. And if we put in things and you don't know how they are being maintained, that causes another problem.
>
> (GV 11/2016)

Reports and complaints of deterioration and vandalism have come up during the first months and years of operations: damaged seats, torn off handgrips, dented vehicle bodies, loose wires and metal rods, turnstile blackouts, littering, inundated stations, broken doors, leaky roofs, etc. The failure to address and repair these shortcomings quickly downgraded the much-lauded high-quality of DART's physical infrastructure. This infrastructural non-linearity of parallel constitution and decay, upgrade and downgrade, demonstrates the mutability of DART, and shows that mess and success are not necessarily contradictory.

A modern daladala?

Operational difficulties such as overcrowding and infrequent bus departures have not decreased since the inauguration of UDA-RT's interim service;

UDA-RT's inconstant service hours and absent timetables are reminiscent of daladala operations. Overcrowding endangers the African BRT model because DART is not seen to improve, but rather to worsen the quality of public transport. Tanzanian newspapers report that passengers want daladala instead of DART (Michael 2018) – quite a different image from the one BRT proponents want to present to the global public (see Chapter 4). Writing in the DART WhatsApp group, Zakai Degera, the UDA-RT service delivery manager, gave vent to his frustration because some drivers still refused to follow the rules of work and the standards of BRT operations after half a year of collecting experience:

> I discovered very, very, very many issues… if we don't work on them by collectively combining our strengths in this city, it will remain being called A MODERN DALADALA and not BRT.[6]
>
> (11/2016)

Bus drivers, ticket vendors and station attendants behaved as they liked or thought best, often due to the lack of clear information and guidelines. They asked themselves: Do we actually need to announce the upcoming stations? Do we need to stop at each station even if the bus is already overcrowded and no one disembarks? When we cannot sell tickets due to a network breakdown, do people get free rides or do we ask them to walk to the next station?

This raises the question of how much DART's current interim services actually differ from daladala operations. Certain practices of DART are surprisingly close to practices of minibuses operating without a written standard or central control. Whereas the Tanzanian World Bank consultant for DART, referring to the physical infrastructure, stressed that DART 'is still the Bogotá model' and 'really the top-class BRT' (JK 10/2016), operational practices have called into doubt whether UDA-RT could meet top-class BRT. Nevertheless, DART has been operational despite the various challenges. During the first weeks of operation, major challenges included the missing installation of the intelligent transport system (ITS) – i.e. activating smart card services, integrating fees for feeder and trunk routes, bus scheduling and fleet management at the control centre – and the partly absent, weak or unstable electricity and internet connection between stations, buses and the control centre. In addition, both UDA-RT and the DART Agency repeatedly problematised the poor communication between the two sides, and further crucial BRT-relevant components were missing: laws and regulations, signs and display indications, ticketing equipment, the DART logo at buses and stations, station furniture like fences, rain gutters and benches. Another initial challenge was the lack of passenger access to certain stations. Keys were missing at some stations, and attendants came late to work, meaning the station could not be serviced.

For the construction period, the construction company mounted signs at each DART station indicating the name of the station and its own logo. When the ISP operations started, the signs stayed because the DART Agency had run out of money for own signs. A couple of months later, the DART Agency asked the construction company to either dismount the signs or to pay a fee to the DART Agency (because the signs function as advertisement for the company). After the company decided to dismount the signs, the DART Agency suddenly offered the company to remount the signs free of charge because the attempt to generate money had failed and the DART Agency preferred those signs over no signs at all. The company willingly remounted the signs. This month, ITDP took some money and replaced them with proper DART signage – shortly before the international *Mobilize* summit is to take place in Dar es Salaam.

(June 2018)

Seen from another perspective, the DART Agency and UDA-RT had the courage and competence to operate DART under precarious conditions, and they were able to solve major challenges within the first weeks of operations. UDA-RT's management did not find this gradual implementation problematic, and passengers apparently did not complain about locked stations – on the contrary, passenger numbers have been growing constantly. Over time, UDA-RT flexibly installed central components like rain gutters and convex mirrors, the AFCS, and implemented feeder services. DART was developing from day to day, and stakeholders evaluated and improved DART continuously: daily visits to the corridor and stations by the operations manager, WhatsApp groups for drivers, station attendants, steering committee, etc. This flexible and gradual implementation provided the opportunity to continuously rediscuss and reassemble DART on various levels, both concerning the operational structure as well as components that needed to be added or adapted.

Looking into detail, DART operates significantly different compared to daladala, especially in terms of the use of intelligent technologies, which require the users of DART to adapt their behaviour. 'People have not gotten used to it' (*watu hawajazoea*), I often heard during the first year of operation. Most passengers had never used such a bus system before, drivers needed to practice manoeuvring an 18-metre articulated bus, mechanics needed to learn how to maintain and repair high-tech buses, passengers had to learn to use turnstiles. From time to time there were misinterpretations, which – again – did not make a stir:

I sit on a DART bus. To my left, a passenger who interestedly observes the window hammer. He takes it out of the holder, and thereby activates the anti-theft device. As the alarm signal shrills, he tries hectically to put the hammer back to the holder, but fails. The bus driver, stopping at the next station, unperturbedly walks to the passenger, takes the hammer and puts it back into the holder. Without comment, the driver walks back to the driver's seat and continues the bus journey.

(September 2016)

I did not observe evidence of the 'feelings of brutal disorientation' that users reported in the initial period of *Transantiago* (Ureta 2015: 95). Perhaps the gradual implementation of DART helped people to get used to the new technology.

Taking a closer look at the gradual implementation of DART's crucial components and how the project's stakeholders organise bus operations, I refer not only to the political background of the intermittent assembling, but also to the functions of these components that exercise social control and discipline urban space (Marquardt 2017; Ureta 2012). Technical objects such as lanes, stations, vehicles, fences, turnstiles and tickets play a role in stabilising DART and reorganising the city in space and time. They contribute to the lived experience of Dar es Salaam's residents, enacting boundaries to separate the new from the existing (Pineda 2010). When humans delegate ethics, duties and values to nonhumans (Latour 1992: 157), these technical objects can become social and political. Power and values are inscribed into these objects, which might create new forms of agency and subjectivity (Von Schnitzler 2008; Winner 1980). How humans interact with an automatic gear, a turnstile or a stop-button is inscribed into the materialities of these objects. DART's technologies 'determine or compel certain actions' (Latour 1992: 225), but not necessarily in the way discipline and control had been inscribed into them. Taking examples of how DART and its technologies are adapted and de-scribed, the following sections discuss DART between the two narratives: the African BRT model versus the so-called modern daladala.

Fluid signs

A result of the gradual implementation of DART was that traffic laws, regulations on the phasing out of daladala, and rules regarding passenger behaviour at stations and on buses were only implemented after the system began operations. Legal arrangements were not dealt with in advance but only came up as incidents occurred. For instance, UDA-RT only started to discuss the size and kind of baggage allowed in the buses after passengers had brought problematic things with them: oversized suitcases, rainwater tanks, car wheels, buckets of fish and big sacks of flour and rice. Even a motorbike and chickens in voluminous cages found their way into DART's buses. Since most people cannot afford a private vehicle, public transport in Dar es Salaam functions as transportation devices not only for persons but also their belongings and merchandise. Especially on the route of Phase 1 – which serviced Dar es Salaam's business centre Kariakoo, the central fish market and ferry terminals to Kigamboni and Zanzibar at Kivukoni – passengers needed to transport their goods. When it became cramped and smelly in the new vehicles, UDA-RT needed to establish an official position: what is allowed as carry-on luggage, what is not, and can UDA-RT charge extra for certain items?

Some general regulations, e.g. regarding traffic laws and customer service, are mentioned in the *BRT Planning Guide* (Wright and Hook 2007, ITDP 2017b), but not these very context-specific issues. Certain regulations had to be created in Dar es Salaam:

> I see a sheet of paper on a wall of Kivukoni Terminal, which is located directly opposite Dar es Salaam's central fish market. The note says: 'You are not allowed to carry fish without putting it into a bucket and closing it with a plastic lid without leakage. The administration.'[7]
>
> (September 2016)

As in this example, various regulations began to appear on sheets of paper, sticky taped to the terminal walls. Such ad hoc notices instructed people: not to use pedestrian bridges for sitting, standing, sleeping, vending or any other activities (*hairuhusiwi kukaa, kusimama, kulala kufanya biashara au shughuli yeyote darajani*); to keep left on pedestrian bridges (*pita kushoto*); to use this counter only if you have the exact fare in cash (*wenye change kamili tu*); to make sure you receive two receipts (*hakikisha unapewa risiti mbili*); and to carry only one piece of baggage with you (*mzigo mmoja tu*). Regulations also stated that transporting live animals is not allowed (*usafirishaji wa wanyama hai* [...] *haruhusiwi*), and that a motorbike does not count as baggage (*pikipiki hayahesabiwi kama mzigo*). Other sheets of paper tell the destination of the bus, which use different parts of the long platform: the section close to the entry point at Morocco Terminal is for buses to Kimara, buses heading for Gerezani service the middle part, and buses to Kivukoni use the rear part – an 'innovative idea' that occurred to UDA-RT, as the company's chief operations officer told me (LH 05/2016).

Material objects are the vehicles of operational control and social values: signs and symbols, colours and letters tell passengers how to behave, which they may either accept or de-scribe. Inside the buses, certain artefacts that were intended to organise and regulate the passenger behaviour caused irritation and misunderstandings. UDA-RT again responded by adding sheets of paper and publicised the 'correct use of UDA-RT's buses' (*matumizi sahihi ndani ya basi la UDA-RT*) in UDA's monthly magazine (UDA 2017). Particularly during the first weeks of operation, much information was provided face to face. Passengers did not hesitate to ask drivers or other passengers where the bus was going to and whether it was an express or a regular line. The system was still new to the passengers; system maps and screens at stations and inside buses had not (yet) been installed as planned:

> At each station, we are going to install smart TV. [...] It will help passengers to get to know information if bus is coming or if route is cancelled. And if it's coming, at what time the bus will be here. That is another stage, you don't see it now, those smart screens. But soon, they will be there.
>
> (AP 09/2016)

This future imaginary, described by the customer operations manager from the AFCS supplier, has not yet been realised. Passenger information has been inconsistent because only some of the buses were equipped with sheets of paper indicating route and destination of the respective bus. During off-peak periods, I experienced how station attendants led the passengers to the correct docking bay and waved the approaching BRT bus nearer. This practice of telling the driver to stop is inspired by daladala practices (see Chapter 1). Automatic announcements have gradually been introduced in the buses, but they remain absent at stations and terminals.

Large technical systems like a BRT system, serving several hundreds of thousand passengers daily, cannot operate efficiently without centrally collected and distributed data, as the CEO of the DART Agency elaborated:

> We have a challenge with the scheduling and dealing with the peak hours, and completing the installation of ITS. Because, apart from the scheduling, I think the crowd control and so on could be easier if we had enough displays of the information system. Like, when a bus stops, [people] scramble for buses because they only see for what has come. They don't know what's coming next or what time it could come. But if there was an information system displaying, saying that: 'The next bus will arrive in two minutes. And another one in five minutes.' So, one wouldn't be scrambling. But right now, no one knows what's there.
>
> (GV 09/2016)

The screens in the buses, which were planned for traffic information and advertisement, have still not been activated, even though UDA-RT staff told me in May 2016 that they would be installed by the end of the month. Nevertheless, (audio-)visual information was guaranteed to a limited extent inside the buses. The next station, terminus, current time and estimated arrival time appeared on simple LED displays. Several bus drivers questioned the need for automatic announcements and electronic displays; according to them, people knew the routes and stops from their previous daladala experience.

Several components inside DART's buses – orange-coloured seats and a pictogram on the one hand, and the yellow *No Standing Area* on the other hand – were not self-explanatory to DART's passengers. First, BRT seeks to cater for passengers with different abilities and needs through measures such as platform-level boarding and seat regulations. Whereas most bus seats are blue, several seats are orange, indicating that they give priority to persons with physical disabilities. Since most of DART's passengers neither understood the meaning of different seat colours nor its pictorial representation of a white person holding a stick in front of a blue background, UDA-RT taped a sheet of paper on the bus window, explaining that these seats are for people with disabilities (*viti hivi ni kwa wenye ulemavu*). Daladala practices do not require signs and written elaborations because passengers follow unwritten social rules that determine whether someone receives priority seating or not. Hence, when users de-scribed the special seats, UDA-RT shifted from pictograms to words. However, the seat's initial script has never been realised. A *YouTube* spot (Bongoflix Star 2018) shows how several young passengers ignore the pictogram until a passenger called 'Mtanzania' (meaning 'Tanzanian' in Kiswahili) hits them with a stick so that they give their seats to an old and a physically disabled person. Mtanzania then explains the buses' intended function, values and regulations to the passengers: 'This vehicle was built and designed under consideration of the needs of all people'.[8] Also the DART Agency (2018) published a video in which two singers explain how to use DART in an appropriate and respectful way. They address the special seats, the dedicated lane and how to behave in situations of overcrowding.

Second, each bus has yellow areas with the red letters NO STANDING AREA: located at the doors to prevent passengers from being squeezed between the hydraulic doors, and located under folding seats in an area dedicated to wheelchairs and baby buggies. The areas are intended to guarantee passengers safety and enforce BRT's values of inclusive, non-discriminatory public transport, respectively. These areas, too, were provided with a sheet of paper – to translate not only the words, but also the meanings and values of why 'it is not allowed to stand here' (*hairuhusiwi kusimama hapa*). When passengers are standing in the door area, the doors should not be able to close even if the driver pushes the door-close button. In addition, buses 'have a mechanism with safeguard to avoid opening of the doors while the vehicle is in motion or that the vehicles moves with open doors' (DART Agency 2014a). Different versions of the initial script of the buses' sensors and doors circulated. An employee of the bus manufacturer told me that light sensors prevent doors from closing when passengers block them. By contrast, service provider staff claimed that there had never been a technology to prevent doors from closing (SC 05/2016; ZD 09/2016). In practice, however, the regulations of the red letters as well as the sheet of paper do not have the intended effect: I witnessed doors closing even though passengers were standing in the yellow area, or even when they were in the process of boarding, with one leg in the bus and the other one on the platform; and buses often set off with open doors, only closing them while the bus was accelerating.

There are at least two reasons for these challenges of translating signs, symbols and letters. Some DART users do not understand either the symbolism or the English language, whereas others are seemingly not interested in conforming to these regulations. In order to stabilise, the new rules of conduct needed to gain acceptance and replace previous (daladala) practices that are primarily based on oral communication between passengers, conductors and drivers. Neither the project's stakeholders nor passengers perceive direct oral communication and sheets of paper as provisional solutions, in contrast to global BRT proponents who complained about missing announcements, station signage and system maps. Also the *BRT Standard* gave DART point deductions for the absence of displays and signage. In Dar es Salaam, electronic audio-visual announcements do not seem to be indispensable. The importance that globally dominant documents and BRT consultants might attach to them is not a given. Hence, announcements and signs that aim at regulating DART's operations are simply absent for most of DART's stakeholders and passengers, but presently absent for BRT planners.

Disciplining drivers, disciplining passengers

In order to support the regulations inscribed into DART's objects, bus drivers and passengers were disciplined in various ways. Not only does signage aim to guide human behaviour, but also the materiality of the DART corridor aims to discipline bus drivers, passengers and other road users. This new space in Dar es Salaam embodies normative politics and traffic laws, and entails disciplinary devices that enforce certain behaviour from its users (c.f. Ureta 2012, 2014): kerbstones and fences, speed bumps and zebra crossings, displays and signs, barcode scanners at turnstiles and electric doors realise new normativities of urban mobility. Technical objects define and stabilise the relationships between actors, and over time they may naturalise and depoliticise these relationships. The political strength of objects is inscribed into their materiality (Akrich 1992: 220–222), which is further transformed to BRT-specific traffic laws – a sometimes cumbersome endeavour.

The corridor of DART symbolises the new that emerges in alternative practices and modes of behaviour on Dar es Salaam's roads. Viewing the city in satellite images or from an approaching plane, the BRT corridor is highly visible, meandering through the city. The corridor, between 20 and 35 metres wide, consists of slabs of 22 centimetres thick ferroconcrete. In the BRT community, concrete is considered to be the best option for road surface because it is more durable than tarmac and thus increases the durability of a BRT system: 'You want it to be maintenance-free as long as possible' (ITDP 2018b). Technocrats highlight that this type of paving is valued for its high quality and durability, and its flexibility and simplicity in terms of repair and maintenance (MH 09/2016; WG 10/2015). Slabs are easily replaceable; a feature that has already proven well in Dar es Salaam due to leaky water pipes (see Chapter 3). The BRT lane is demarcated in two ways from the mixed-traffic lanes: physically by its high kerbstone, and visually by its different colour. It is almost impossible to get onto the corridor due to fences near BRT stations and the kerbstone along the corridor – the 'crash barrier that prevents motorists from taking advantage of its free-flowing smoothness' (The Citizen Tanzania 2016).

Despite its material strength, enforcing BRT law on the corridor had been challenging. Drivers of daladala, cars, motorbikes, auto-rickshaws and bicycles had to adapt to new logics of movement in the city. New practices coming along with the corridor embody numerous transitions that entailed tensions and conflicts. Particularly the first two years of the ISP operations were accompanied by complaints about private cars and motorbikes invading the dedicated BRT lane. Unauthorised vehicles would sometimes sneak onto the corridor at junctions where the kerbstone is interrupted to avoid congestion. They entered the BRT sphere either unintentionally or because they were used to vehicles being tolerated on the corridor during the construction period. Hence, new traffic laws had to be passed to guarantee the sole rights of usage for DART buses on the dedicated BRT lane. Directly before opening the system, the bus operator announced on public media that no other vehicle, including government and police, was allowed to use the BRT corridor anymore because UDA-RT needed to guarantee uncongested BRT services. One month after the system's launch, a police car entered the corridor and collided with a BRT bus. People guessed that the driver of the police car wanted to bypass the traffic on the mixed-traffic lane and expected BRT services to give way to them. (Even if the bus driver had wanted to give way, they could not have because of the physical arrangement of the kerbstone.) Even though DART's stakeholders took measures, e.g. by publicly denouncing those who performed 'bad road usage' (*matumizi mabaya ya barabara*) of the corridor with their full name and a photo in the DART WhatsApp group, a clear regulatory framework was long missing. Zakai Degera announced in the chat group:

> With this announcement and starting from today, Thursday 29/03/2018 there is no car that is allowed to drive on this road (BRT LANES) [...] LEADERS PLEASE HELP US BRT IS BEING INVADED PEOPLE JUST DRIVE ON IT NOW.[9]

Because the use of a dedicated lane is one of main features that distinguishes BRT from regular bus systems, the slow and patchy enforcement of the rules on DART's corridors threw its legitimacy as a genuine BRT into question. In addition, it shows that deviations from the original plans were stabilising while the initial plan to enforce the dedicated lane was losing prominence – at least temporarily.

Also, pedestrian crossings and the legal enforcement of pedestrians' right-of-way are central BRT features that illustrate the changing practices directly resulting from the transitional process. Pedestrians are only allowed to cross the corridor using pedestrian bridges at terminals or zebra crossings near stations, a shift from previous practice: neither pedestrians nor drivers

Figure 6.3 Pedestrian crossing with a speed bump at Jangwani station (author's photo 05/2016).

in the mixed-traffic lane used to take any notice of the white stripes on the ground. But BRT drivers have had to take notice from the start because they were instructed to do so in their BRT driving training. In the first weeks of operations, pedestrians waiting at zebra crossings for the bus to pass were surprised to see massive vehicles full of passengers brake to let them cross. Pedestrians started to accept this new bus–pedestrian relationship, and to expect the same behaviour from other vehicles (see Figure 6.3). Successively, different groups of road users have grown used to acting according to the inscribed behaviour of pedestrian crossings. Nevertheless, many private drivers do not follow the script, meaning that pedestrians cannot blindly follow the script and must instead assess whether vehicles are going to stop for them.

In addition to the white stripes, the DART station zebra crossings have speed bumps that force regular vehicles to decelerate. Two conceptualisations help understand the effect of this material component on DART's usage: First, Ureta describes speed bumps as heterogeneous and performative 'sociotechnical devices designed with the explicit aim of disciplining users' bodies in accordance with certain predetermined programs or plans' (2012: 596). Drawing on Foucault's conceptualisation of discipline (1979), disciplinary devices encode scripts so that 'the abnormal components start self-governing; materially enforcing their alignment' (Ureta 2014: 371–372). Forms of graphical information like advertisements and signage, as well as physical components like fences and turnstiles, act as a self-governing power that discipline the users' bodies of Santiago de Chile's metro system

(Ureta 2012: 606; Ureta 2015). As a disciplinary device, the speed bump regulates and disciplines the users of the corridor and becomes an actor itself that materialises normative policies (see also Gandy 2002; Winner 1980). Second, Latour also uses the speed bump to illustrate technical action, which he refers to as technical mediation or technical delegation (1994, 1999). A speed bump, also called *gendarme couché* or sleeping policeman, acts as a mediating technology. Latour jokes about how a cop changes into a barrel of wet concrete – 'the characteristics of policemen become speed bumps'; speed bumps are full of engineers, lawmakers, etc. (1999: 190). As a mediating technology, a speed bump disciplines the drivers by changing their behaviour in a forceful way. Differently than a sign that tries to make the driver to slow down through morality and reflection, the speed bump does so by triggering pure selfishness and reflex action: 'The driver's goal is translated, by means of the speed bump, from "slow down so as not to endanger students" into "slow down and protect your car's suspension"' (Latour 1999: 186). Thus, the programme of action to slow down vehicles has been translated from signs to concrete and pavement. Speed bumps are non-negotiable, in contrast to signs or police officers. In Dar es Salaam, the design of the zebra crossing realised their goal: white lines on the corridor for educated BRT drivers and speed bumps with white lines for mixed-traffic users led to discipline, which enforced the priority of pedestrians and thereby realised one of BRT's core ideals.

Other means to discipline drivers and passengers were quickly left behind. For instance, DART's drivers had learned to recite a greeting that informs passengers how to behave in a BRT bus:

> Good afternoon, dear passengers. My name is Jason. I am your driver. Announcement: those passengers who are standing, I beg you not to stand in the yellow marked door area because our doors open with electric sensors. I also ask you not to throw your waste on the floor. Put the waste into the dustbin. Thank you very much. Have a nice trip.[10]

In the early days of operations, drivers tried to adhere to the comfort and safety policies they had learned. But practices soon changed due to resistance against it, time constraints and a subjective lack of necessity. Because drivers and hydraulic doors did not comply with this comfort and safety script, a gap appeared between the initial prescription and the actual subscription (Akrich 1992; Latour 1992). A similar change in the script occurred regarding a technology designed to prevent drivers from speeding, which releases a high-pitched beeping sound if the bus exceeds 50 km/h. However, in practice many drivers speed in some sections of the corridor and simply ignore the beeping. Hence, they do not adapt their driving to the initial script. The CEO of the DART Agency joked that speeding drivers perceive this sound as music rather than as a warning signal. DART's drivers seem to be more resistant to this disciplining script than Latour, who could stand the alarm of his unfastened seat belt for only twenty seconds (1992: 225). In Dar es Salaam, the beeping is no longer a guarantee for maximum speed, as it was supposed to be in the initial script.

Gradually implementing DART's physical infrastructure and its disciplinary devices has involved further questions of responsibility and reliability, ownership and the right to participate. Because the ISP began operations without fencing off the terminal platforms and bus lanes from public road space, passengers entered the terminals from the side, circumventing the turnstiles (see Figure 6.4). UDA-RT disliked this practice for two reasons. First, the company lost revenue due to reduced ticket sales. Second, passengers risked being run over by a bus. Because it had to make money quickly to pay back its various loans, UDT-RT prioritised enclosed stations and a functioning ticket system when it began operations. As the owner of the terminals, the Tanzanian government was responsible for constructing structures that guaranteed safe and smooth operations. However, as construction works had already exceeded the expected costs, both TANROADS and the DART Agency agreed with UDA-RT's plan to construct a simple temporary solution, whose costs the DART Agency would reimburse. As the CEO of UDA-RT elaborated: 'We will basically do a chicken wire fence around.

Figure 6.4 Kimara terminal without passenger barriers during first days of operation (author's photo 05/2016).

I am quite sure that's enough to make sure that people don't get into the buses without paying' (HG 05/2016). A couple of days later, however, UDA-RT management changed its mind:

> What happened at Kimara is: we made a very solid fence. And TANROADS came in and said: 'No, this is not what we said. It should be a chicken wire'. We made much more solid than what they wanted. To me, he doesn't make sense. And I put my foot down, I am saying: 'You allow us to finish like it is, or nothing goes on'.
>
> (HG 09/2016)

The simple solution, the chicken wire, has never materialised. DART's terminals are either protected by the 'very solid fence' or by ropes that UDA-RT wrapped around the pillars after losing hope that TANROADS or the DART Agency would reimburse the company for the fences already installed. This case recalls the simple ticketing solution that UDA-RT ignored and established a 'very solid', long-term solution for ticketing, on the one hand, and the provisional daladala physical infrastructure, on the other hand.

Twisted ticketing and timing

Apart from the dedicated lanes and the passenger capacity of the buses, ticketing is a major difference between daladala and DART, evoking new relations and interactions. Central stakeholders, including UDA-RT and the steering committee, described ticketing as the major issue of DART's implementation. The infrastructure at stations and terminals cannot cope with the high ridership at peak hour, the partially weak electronic and Wi-Fi network that takes a long time to validate tickets, the comparatively high number of passengers using paper tickets, and the odd fare of TZS 650. The ITS has various functions, most importantly selling tickets and calculating the service demand in real-time. Turnstiles require passengers to scan a valid paper ticket or a smart card. Passengers needed to learn to understand how this new technology works. Since most passengers initially had difficulties scanning their tickets and smart cards, station attendants mediated in this passenger–turnstile relationship. Station attendants took over the scanning and pushed the turnstile for the passengers so that they could pass. The half-installed AFCS and smart cards brought various technopolitical challenges with them, materialising in non-linear implementation steps, as well as delays and continuous operational changes.[11] Smart cards were sold for only half a day before UDA-RT returned to paper tickets for over a month; network connection problems have been (re)appearing from time to time; stations sometimes run out of paper rolls and change – even after several years of operation; during peak hours, congestion relocated from the road to the ticket booths at station entrances (see Figure 6.5).

Figure 6.5 People lining up for buying a ticket at a DART station (author's photo 05/2016).

The eventual introduction of smart cards opened a new world to DART's passengers and partially changed their behaviour; it raised the operational standard of UDA-RT and saved money for everyone connected to the smart card universe. Only after smart cards were implemented could feeder services from Kimara to Mbezi begin. However, after less than six months, UDA-RT's supply of 200,000 smart cards ran out, forcing that everyone that had not yet bought one to use paper tickets.

> I meet a friend from Dar es Salaam in a bar in Frankfurt. When the subject of DART cropped up, he proudly presents me his DART smart card, which have become a limited good in Dar es Salaam, being even sold on the black market. I remember how the corporate communication manager of SUMATRA was sceptical about the acceptance of smart cards in Dar es Salaam. When I met him two years ago, he described Tanzania as a 'cash society' that would neither be willing to pay a deposit for a smart card, nor to transfer money for several trips on it. It turns out that his prediction was wrong.
>
> (March 2018)

Passengers without smart cards experience significant disadvantages in the ticketing system, particularly feeder users, because the reduced transfer fare of TZS 800 from trunk to feeder and vice versa is only available with a smart card. Passengers without a smart card need to buy two separate tickets of TZS 650 for trunk and another one of TZS 400 for feeder. In addition, because feeder stops do not have ticket booths, a smart card is needed to begin a journey from a feeder stop.

UDA-RT employees and DART users have adopted a variety of strategies to deal not only with the lack of smart cards, but also cases where there is no change at ticket booths, or absent network connections at stations and terminals. The high number of staff at the stations despite the complex AFCS, which normally replaces the human workforce, was helpful in these situations.[12] In addition, UDA-RT mounted sheets of paper at terminal ticket booths, dividing customers into three categories: those with the exact fare (*wenye change kamili tu*), students (*wanafunzi*), and the remaining customers (*normal*). At some stations with only one ticket booth, ticket vendors had written on a sheet of paper: 'We kindly ask for the exact TZS 650, or you might also put 150 on top' (*Tunaomba 650 TSh kamili au weke 150 juu*) – meaning that customers would receive exactly TZS 500 in change when they gave TZS 1,150 for a single ticket that costs TZS 650. When station attendants were short of change and passengers could not pay the exact fare, the ticket price could temporarily rise from TZS 650 to 700. In addition, freelance money changers set up in front of the ticket booths, offering what they typically offer to daladala conductors who run out of change: a stack of nine TZS 100 coins in exchange for a TZS 1,000 banknote. However, UDA-RT promptly prohibited this business, arguing that it did not correspond with the idea of a modern transport system. At other stations, ticket vendors did work-sharing to supply passengers with tickets more quickly: in the ticket booth, one vendor sold tickets to passengers who needed change, while in front of the ticket booth a second vendor sold pre-printed tickets to customers who had the exact change. Surrounded by impatient customers, the vendor hectically handed out tickets and shoved the money into their pockets that served as a temporary cash register. Also, during power cuts, staff were much more helpful than non-functional machines. On a Saturday night around eight p.m., a power cut sent Kivukoni Terminal into complete darkness. Only a small generator provided electricity for two ticket vending machines, so that ticket sales were secured. Vendors used their phone lights to see. Since the turnstiles were not working, everyone had to buy paper tickets, whose validation date and time were checked by station staff at the terminal's manual gate. The flexibility and creativity in these different situations enabled UDA-RT to operate despite the absence of supposedly crucial components. Even though global BRT technocrats do not welcome this improvised character, the system has proved to be able to (temporarily) function without central technological components while still retaining a certain level of quality of service.

Whereby the AFCS without smart cards worked on DART's trunk route from the beginning, feeder operations were initially intermittent, stopping entirely for several weeks after less than a day of services because smart cards were needed on feeder routes. Even after one (out of eleven) feeder route finally became operational, feeder ticketing remained provisional for several months:

> Three ticket vendors sit on a bench at Mbezi Luis feeder terminal. They carry IDs from the fare collection company, but they have neither uniforms nor a ticket booth. One vendor sells and tops up smart cards with a mobile ticket vending machine. Another one sells paper tickets, which they had pre-printed. He uses a simple plastic bag as cash register. The third one provides information about tickets options, fares and route network to the customers. Inside the bus, passengers either scan their smart cards at the validator, or they hand their paper ticket to the bus attendant who then scans the pile of tickets during the journey; a measure to fasten the buses' departure.
>
> (September 2016)

DART feeder services are characterised by several shapes of gradual becoming and multiple, intermittent temporalities. Numerous practices of the feeder system resemble daladala more than BRT. The notion of 'modern daladala' serves best for the feeder, which is an amalgam of global BRT ideals, DART operations and daladala practices. The passenger-based access fee planned for DART's feeder routes (in contrast to an access fee based on kilometres scheduled, typically applied on trunk routes) follows similar logics to daladala operations (DART Agency 2014b: 44; see also Chapter 5). Daladala-like practices are not necessarily a weakness or the result of misconduct by the bus operator. Since feeder routes do not have the privilege of a dedicated lane, they are slow (see Figure 6.6). Transport planners underestimated the high volume of traffic on this mixed-traffic road, which brings upcountry buses to Ubungo bus terminal and lorries to Dar es Salaam port. A journey that was calculated to take 15 minutes can easily take twice as long, and thus scheduling can be quite inaccurate. In addition, the feeder buses' relatively low chassis can scrape the ground on speed bumps along the road. The speed bumps' inscribed discipline thus not only leads to less speeding of regular vehicles but also decreases the BRT's speed and consequently its quality of service. Following daladala practices, drivers and bus attendants have developed a strategy to compensate for the lack of bus priority and to save time: they only halt at a feeder stop if passengers explicitly want to get on or off the bus. Thus, passengers on the bus must push the stop-button (which is not necessary on the trunk corridor), or bus drivers hoot when approaching a feeder stop, expecting waiting passengers to give them a hand signal to stop. Such practices are reminiscent of daladala, where the conductor gives the driver a signal by shouting 'drop off' (*acha*) or by knocking on metal with a coin.[13] Hence, the villager (*wakijiji*) reaching

Figure 6.6 Mixed traffic on the feeder route between Mbezi Luis and Kimara (author's photo 09/2016).

out to hail a DART bus would not have behaved inappropriately on a DART feeder stop, only on a trunk service (see Chapter 1).

Learning to control

The last example to demonstrate the astonishing levels of operation achievable without having everything in place at the outset concerns the gradual implementation of DART's control centre. A control room is a 'key site that enables the city's infrastructural life by securing urban flows and maintaining the city's circulations' (Luque-Ayala and Marvin 2016: 196). ITDP describes a BRT's control centre as the 'brain that controls the nervous system' (ITDP 2018a). Its core function is to guarantee a level of service quality by pursuing the operational schedule and to calculate the distribution of revenue between operators and the national government. UDA-RT's chief operations officer told me that the ITS knows the schedule, the actual position of each bus and how many passengers are in the system in real time (AP 09/2016). A bus operations specialist illustrated:

> From the control room, we can monitor the service level of the buses, arrival and departure time, waiting time at the stations. So, we can monitor it. But as long as this control room is not ready, we will do

the manual calculations. [...] On a larger scale, you need some good software, which will do the all-up management, the depot management, the bus management, the crew and staff management. A lot of things become part of the scheduling.

(AM 09/2015)

DART's stakeholders were able to supply bus services despite the absence of presumable indispensable technological components, including almost all technical devices and services of the ITS and the control centre. It took almost a year to release the AFCS equipment from customs. Thus, UDA-RT's employees only began to receive training in how to control, supervise and coordinate the operations shortly before operations began. In May 2016, the control centre was virtually non-existent: located in the head office of UDA-RT at Jangwani Depot, desks were empty apart from a few personal computers. No communication was possible between the control centre, buses and stations because the fibre connection was still under construction:

> The operations manager is desperately waiting for the control centre to be installed so that he can transfer the calculation of fuel consumption and the organisation of bus inspections from manual to automatic. He questions the whole DART system and argues that BRT is only worth the label when the system works under central analysis and control. Fundamental questions were still under discussion: can UDA-RT actually be the owner and operator of the AFCS, who has access to the data, and when will the control centre be handed over to the DART Agency?
>
> (May 2016)

Over the course of the first year of ISP operations, the control centre was partially installed. It gained a more material presence and the network could be largely stabilised, but staff was still rarely present, and screens are often turned off:

> DART's stakeholders are frustrated that installation and training takes enormously longer than expected. Data is there but they are not able yet to read the data, to give meaning to the numbers. But at least they seem to be quite excited about the new technology: 'What we see is exactly what is happening on the road,' says UDA-RT's chief operations officer.
>
> (September 2016)

> Finally! Buses appear with their real-time location on the screens. At the transport conference in Nairobi, the CEO of the DART Agency happily shares this progress with his East African colleagues, illustrated by photos: 'This is the control room and some screens where

we get real-time movement of the buses, you can see them. But as I said, it's not yet complete but you can see the buses. And it even has the dotted colours: if it's red, it tells you that bus is late; if it's green, that bus is on time. If it's orange it's a bit late. That kind of information, we're enjoying it.' The infrastructure has been fully installed and will gradually get used: bus tracking, scheduling and communication with the buses, all done from the corner of UDA-RT's main office that functions as DART's control centre. However, the CEO of the DART Agency complains that the government is given only limited access to the data.

(November 2016)

The control centre brought new concerns to Dar es Salaam: who governs the data, who distributes the revenue and who decides about the supply of buses on the corridor? Data have become valuable because it is transferable in money and power. The prestigious BRT control centre of Ahmedabad that the Indian operations specialist Rachit Padmanabhan had installed before he came to Dar es Salaam is not comparable to DART's control centre. His approach and his expectations differed considerably from those of UDA-RT. His job was to install scheduling, fare collection management and bus fleet management. The employees of the Tanzanian transport company usually operate without computer programmes, irritating global consultants and transport specialists (AE 05/2016; RP 09/2016). Thus far, bus scheduling had been done using hand-written tables. The CEO of UDA-RT was convinced that the complex AFCS is not necessary, and he argued that when he inspects the stations, his eyes tell him whether more buses were needed to secure adequate operations (HG 05/2016). The Indian specialist reasoned from this 'Tanzanian' attitude that it was significantly easier to work in India because Indians were more open to technology and transport planning. He felt that no one at UDA-RT was interested in his calculations, sheets and plans, whereby he thought in programmes and spreadsheets. Even though he personally inspected the BRT infrastructure daily, perceiving with the own eyes became secondary once technology was involved:

In the second week of operations, Rachit explains me his spreadsheets: what is happening 'out there right now' – as if the numbers on the screen were the reality, determining where buses are. He says that passengers cannot be right when they complain about unpunctual and irregular departures. He points at his spreadsheet: 'You see, one bus starts from there at 03.05 hours, the next one at 03.09 hours.' In fact, the tables are solely based on his calculations and not on the actual positions of the buses, because GPS is yet to be activated in the buses.

(May 2016)

The absence or inactivity of the control centre led to demand-based operations, which have probably served the customers better in certain respects than if UDA-RT had followed a ready-made operational plan. UDA-RT extended the service hours and routes in stages and added buses to corridors in response to the increasing demand. The flexible handling of this challenging situation, growing out of a major deficit, does not only have disadvantages such as inconsistency of services; the continuous opportunities to modify also imply an openness to the system, which tends to result in a general improvement and stabilisation of the system in the long term. This approach is typical of the general, flexible formation of DART: learning by doing.

The operational changes have paralleled the fate of the ISP itself: just as UDA-RT has developed from an interim service provider to a permanent service provider, so too have DART operations become a permanent interim assemblage. The gradual implementation has facilitated continuously (re)making DART. Stabilisation stays flexible and negotiable on the one hand, while short-term solutions become permanent on the other hand. The ad-hoc and reactive manner of disciplining and stabilising DART has happened not only from plan to practice, but also within practice itself.

The various changes and adjustments of DART demonstrate that an infrastructural process is not linear but multiple and complex, and that temporalities and spatialities are not separable from each other (Bowker 2015; Thrift and May 2001). Present absences and twisted temporalities are the expression of the new technological system's adaptation and the continuous negotiations on matters of power and participation. Conceptualising DART as an emerging, highly mutable technopolitical assemblage (Collier and Ong 2005; Hecht 2011) that consist of changing scripts (Akrich and Latour 1992; Ureta 2015) allows to perceive the gradual and multiple, reverse and opposed temporalities of the infrastructural project, as well as to distinguish between the shiny external view of DART and how DART actually works on the ground. DART is not only emerging; it is also (prematurely) deteriorating and ageing. The system resembles daladala in various regards, and the image as a modern daladala has been re-inscribed into the assembling, so that a clear differentiation between the global BRT model and so-called 'paratransit' services is thrown into question. DART's material components and practices express the challenges of stabilising and disciplining its operations. These examples not only provide insights into technopolitical decision making and negotiations of power, but they also explicate how plans of best practice BRT, designs of DART and context-specific understandings intermingle to produce the fluid formation of BRT in Dar es Salaam.

Notes

1 Articulated buses have a length of around 18 metres and rigid buses have a length of around 12 metres.

2 *Tungependa kama angeweza kuja wakati tumeshakuwa na smooth operations, tume-shasolved most of the challenges. [...] Hii inauguration itafanyika, lakini when sasa?*
3 *Also ich muss fairerweise sagen: wenn man vorbeifährt und man würde nicht als Ingenieur darauf gucken, sieht man zwar, dass irgendwas verkehrt ist. Aber die Details sehen Sie nur als Ingenieur. [...] In Deutschland würden Sie es nicht abgenommen kriegen, weil einfach auch die Struktur beschädigt wurde. Das heißt: die ganze Station ist nach rechts geruckt und irgendwann fragt man sich dann, ob die Statik das noch mitmacht.*
4 *UDA-RT mnaombwa muwaelimishe madereva [...] kuwa kituo cha BARUTI hakitumika, ni kibovu. Waache kushusha abiria kwani ni hatari kwa usalama wa abiria. Wanawashusha kwa nje ya kituo at raised platform then abiria wanaruka chini na kuvuka barabara sehem hatarishi kwao.*
5 *Das Prinzip ist wirklich: der Unternehmer baut nach dem Plan, den der Ingenieur zum Bauen freigegeben hat. [...] Die existierende Brücke sollte in jedem Fall nur verbreitert werden. Technisch richtig wäre gewesen, sie abzureißen und sie höher zu legen. Was natürlich wieder zu einer Kostenexplosion geführt hätte. Und dadurch, dass das Design auch für das Busdepot schon auf die Höhenlage der existierenden Brücke abgestimmt war, hat man halt in der Sequenz gesagt: „Wenn das nicht, dann das auch nicht, und das auch nicht."*
6 *Nimegundua mambo mengi sana sana sana... Ambayo kama tusipoyafanyia kazi kwa kuunganisha nguvu pamoja basi BRT kwenye Jiji letu itabakia kuitwa A MODERN DALADALA na siyo BRT.*
7 *Huruhusiwi kubeba samaki bila kuweka kwenye ndoo, na kufunika na mfuniko wa plastic usiovuja. Utawala.*
8 *Gari hili limetengenezwa na kudesigniwa kwa kuangalia hali za watu wote.*
9 *Kwa taarifa hii na kuanzia leo tarehe 29/03/2018 Siku ya Alhamis hakuna Gari yoy-ote inayoruhusiwa kupita kwenye Barabara hizo (BRT LANES). [...] WAKUBWA TUSAIDIENI BRT IMEVAMIWA WATU WANAJIPITIA TU SASA.*
10 *Habarini za mchana abiria wangu. Kwa jina, naitwa Jason. Mimi ni dereva wenu. Angalizo: abiria wote waliosimama, naomba msisimame kwenye milango na alama ya manjano kwani mlango wetu unafungua kwa sensor ya umeme. Pia, abiria usi-tupe takataka nje ovyo. Weka takataka kwenye chombo maalum ndani ya basi, yaani dustbin. Asanteni sana. Karibuni.*
11 As of February 2020, the AFCS was still not fully installed (Malanga 2020).
12 In addition, the use of humans increases the level of independence from inter-national expertise and spare parts, as the CEO of the DART Agency explained to his colleagues from the neighbouring countries at the transport conference in Nairobi: 'We need our own experts. Train your people early so that they can run the ITS system [and] don't need to call Belgium. [...] People think, scheduling it easy but that's not true'.
13 D'hondt describes the drop-off routine in the 'low-tech environment' of a da-ladala (2009: 1962) as free from mediating technologies because passengers and conductors can only communicate amongst themselves and not with other actors in remote locations. Following this argument, communication differs significantly between daladala and DART, at least in theory. The BRT system's GPS, displays and wires enable passengers and drivers to communicate through the inclusion of a variety of other actors, most importantly the programmes and employees in the control centre, who calculate the current passenger load and traffic demand.

References

Akrich, M. 1992. "The De-Scription of Technical Objects". In: W. Bijker and J. Law (eds.), *Shaping Technology/Building Society. Studies in Sociotechnical Change*. Cambridge, MA: MIT Press, 205–224.

Akrich, M. and B. Latour 1992. "A Summary of a Convenient Vocabulary for the Semiotics of Human and Nonhuman Assemblies". In: W. Bijker and J. Law (eds.), *Shaping Technology/Building Society. Studies in Sociotechnical Change.* Cambridge, MA: MIT Press, 259–264.

Barry, A. 2015. "Infrastructural Times". In: H. Appel, N. Anand and A. Gupta (eds.), *The Infrastructure Toolbox.* Fieldsights, September 24. Accessed at *Cultural Anthropology* (https://culanth.org/fieldsights/series/the-infrastructure-toolbox). Published 24.09.2015, retrieved 17.04.2020.

Bongoflix Star 2018. "Vituko vya Mwendokasi – Bakora TV Show". Accessed at *YouTube* (https://www.youtube.com/watch?v=_DzLiQnsrcU&frags=pl%2Cwn). Published 24.05.2018, retrieved 11.11.2018.

Bowker, G. 2015. "Temporality". In: H. Appel, N. Anand and A. Gupta (eds.), *The Infrastructure Toolbox.* Fieldsights, September 24. Accessed at *Cultural Anthropology* (https://culanth.org/fieldsights/series/the-infrastructure-toolbox). Published 24.09.2015, retrieved 17.04.2020.

Callon, M and J. Law 2004. "Introduction: Absence – Presence, Circulation, and Encountering in Complex Space". *Environment and Planning D* 22:1, 3–11.

Collier, S. and A. Ong 2005. "Global Assemblages, Anthropological Problems". In: A. Ong and S. Collier (eds.), *Global Assemblages. Technology, Politics, and Ethics as Anthropological Problems.* Malden, MA: Blackwell, 3–21.

Collier, S. 2006. "Global Assemblages". *Theory, Culture & Society* 23, 399–401.

Cresswell, T. 2012. "Mobilities II. Still". *Progress in Human Geography* 36:5, 645–653.

D'hondt, S. 2009. "Calling the Stops in a Dar-es-Salaam Minibus. Embodied Understandings of Place in a Drop-off Routine". *Journal of Pragmatics* 41, 1962–1976.

DART Agency 2018. "DART Documentary". Accessed at *DART Agency* (https://www.dart.go.tz/galleries/listing/videos). Published 08.09.2018, retrieved 17.04.2020.

DART Agency 2017. "DART Galleries". Accessed at *DART Agency* (https://www.dart.go.tz/galleries/listing/photos/0). Published January 2017, retrieved 17.04.2020.

DART Agency 2015. "Interim Service Provider Agreement for the Provision of Interim Bus Transport Service for Phase 1 of the DART System, Dar es Salaam, Tanzania". Executive Copy, April 2015. (*unpublished*).

DART Agency 2014a. "Bus Output Specifications". Updated by Transaction Advisory Team. Version 3.2, June 2014. Dar es Salaam. (*unpublished*).

DART Agency 2014b. "Dar Rapid Transit (DART). Project Phase 1. Project Information Memorandum". Final Version, May 2014. Dar es Salaam. (*unpublished*).

DART Agency 2014c. "DART Souvenir on Marketing Consultation Meet on Phase One of the DART Project. Bid to Make a Difference". Accessed at *DART Agency* (http://dart.go.tz/en). Published June 2014, retrieved 20.07.2016.

De Laet, M. and A. Mol 2000. "The Zimbabwe Bush Pump. Mechanics of a Fluid Technology". *Social Studies of Science* 30:2, 225–263.

Foucault, Michel. 1979 (1975). *Discipline and Punish. The Birth of the Prison.* Harmondsworth: Penguin.

Frers, L. 2013 "The Matter of Absence". *Cultural Geographies* 20:4, 431–445.

Gandy, M. 2002. *Concrete and Clay. Reworking Nature in New York City.* Cambridge, MA: MIT Press.

Giliard, S. 2017. "DART Scoops Award, Admits Limitations". Accessed at *Daily News Tanzania* (https://www.dailynews.co.tz/news/dart-scoops-award-admits-limitations.aspx). Published 18.07.2017, retrieved 17.04.2020.

Graham, S. and N. Thrift 2007. "Out of Order. Understanding Repair and Maintenance". *Theory, Culture & Society* 24:3, 1–25.

Graham, S. and S. Marvin 2001. *Splintering Urbanism. Networked Infrastructures, Technological Mobilities and the Urban Condition.* London, New York: Routledge.

Harris, A. 2013. "Concrete Mobilities. Assembling Global Mumbai through Transport Infrastructure". *City* 17:3, 343–360.

Harvey, P. and H. Knox 2015. *Roads. An Anthropology of Infrastructure and Expertise.* Ithaca, London: Cornell University Press.

Harvey, P. 2010. "Cementing Relations. The Materiality of Roads and Public Spaces in Provincial Peru". *Social Analysis* 54:2, 28–46.

Hecht, G. 2011. "Introduction". In: ibid. (ed.), *Entangled Geographies. Empire and Technopolitics in the Global Cold War.* Cambridge, MA: MIT Press, 1–12.

Hetherington, K. 2004. "Secondhandedness. Consumption, Disposal, and Absent Presence". *Environment and Planning D* 22:1, 157–173.

Howe, C., J. Lockrem, H. Appel, E. Hackett, D. Boyer, R. Hall, M. Schneider-Mayerson, A. Pope, A. Gupta, E. Rodwell, A. Ballestero, T. Durbin, F. el-Dahdah, E. Long and C. Mody 2015. "Paradoxical Infrastructures. Ruins, Retrofit, and Risk". *Science, Technology, & Human Values* 41:3, 1–19.

ITDP 2018a. "BRT Planning 101". Accessed at *ITDP* (https://www.itdp.org/publication/webinar-brt-planning-101/). Published 02.02.2018, retrieved 17.04.2020.

ITDP 2018b. "BRT Planning 401: Infrastructure and Design". Accessed at *ITDP* (https://www.itdp.org/2018/05/04/webinar-brt-planning-401/). Published 04.05.2018, retrieved 17.04.2020.

ITDP 2017a. "Santiago, Chile Putting Pedestrians First". *Sustainable Transport*, Winter 2017, No. 28. Accessed at *ITDP* (https://www.itdp.org/category/content-type/publication/magazine/). Published 02.02.2017, retrieved 11.11.2018.

ITDP 2017b. "The Online BRT Planning Guide". 4th Edition. Accessed at *ITDP* (https://brtguide.itdp.org). Published 16.11.2017, retrieved 17.04.2020.

ITV Tanzania 2016. "Rais Magufuli achukizwa na uharibifu na matumizi mabaya ya miundombinu ya DART" Accessed at *YouTube* (https://www.youtube.com/watch?v=P68yYqGhtKA). Published 02.06.2016, retrieved 17.04.2020.

Jackson, S. 2015. "Repair". In: H. Appel, N. Anand and A. Gupta (eds.), *The Infrastructure Toolbox.* Fieldsights, September 24. Accessed at *Cultural Anthropology* (https://culanth.org/fieldsights/series/the-infrastructure-toolbox). Published 24.09.2015, retrieved 17.04.2020.

Jackson, S. 2014. "Rethinking Repair". In: T. Gillespie, P. Boczkowski and K. Foot (eds.), *Media Technologies. Essays on Communication, Materiality, and Society.* Cambridge, MA: MIT Press, 221–240.

Jacobsen, M. 2016. "Temporalities of Assembling Transport Systems. Presences and Absences in an Intermittent Process". In: J. Mewes and E. Sørensen (eds.), *Ethnographies of Objects in Science and Technology Studies.* Ruhr-University Bochum, 52–64.

Jumanne, S. 2019. "Can Roads/Buildings at Flood-Prone Jangwani Be Elevated?" Accessed at *The Citizen Tanzania* (https://www.thecitizen.co.tz/news/-Can-roads-buildings-at-flood-prone-Jangwani-be-elevated-/1840340-5122606-gwdukgz/index.html). Published 19.05.2019, retrieved 17.04.2020.

Ka'bange, A., D. Mfinanga and E. Hema 2014. "Paradoxes of Establishing Mass Rapid Transit Systems in African Cities. A Case of Dar es Salaam Rapid Transit (DART) System, Tanzania". *Research in Transportation Economics* 48, 176–183.

Larkin, B. 2008. *Signal and Noise. Media, Infrastructure, and Urban Culture in Nigeria.* Durham, NC: Duke University Press.

Latour, B. 2005. *Reassembling the Social. An Introduction to Actor-Network-Theory.* Oxford: Oxford University Press.

Latour, B. 1999. *Pandora's Hope. Essays on the Reality of Science Studies.* Cambridge, MA: Harvard University Press.

Latour, B. 1996. *Aramis or the Love of Technology.* Cambridge, MA: Harvard University Press.

Latour, B. 1994: "On Technical Mediation". *Common Knowledge* 3:2, 29–64.

Latour, B. 1992. "Where are the Missing Masses? The Sociology of a Few Mundane Artifacts". In W. Bijker and J. Law (eds.), *Shaping Technology/Building Society. Studies in Sociotechnical Change.* Cambridge, MA: MIT Press, 225–258.

Latour, B. 1986: "Visualisation and Cognition. Drawing Things Together". *Knowledge and Society – Studies in the Sociology of Culture Past and Present* 6, 1–40.

Luque-Ayala, A. and S. Marvin 2016. "The Maintenance of Urban Circulation. An Operational Logic of Infrastructural Control". *Environment and Planning D* 34:2, 191–208.

Malanga, A. 2020. "Fresh Move to Sort Out Dart Chaos". Accessed at *The Citizen Tanzania* (https://www.thecitizen.co.tz/news/1840340-5462526-af1mr4/index. html). Published 20.02.2020, retrieved 17.04.2020.

Marquardt, N. 2017. "Zonen infrastruktureller Entkopplung. Urbane Prekarität und soziotechnische Verknüpfungen im öffentlichen Raum". In: M. Flitner, J. Lossau and A. L. Müller (eds.), *Infrastrukturen der Stadt.* Wiesbaden: Springer VS, 89–109.

Masare, A. 2017. "Tanzania Eyes Sh2.5 Trillion World Bank Lending". Accessed at *The Citizen Tanzania* (http://www.thecitizen.co.tz/News/Tanzania-eyes-Sh2-5-trillion-World-Bank-lending/1840340-3787670-format-xhtml-bile9az/index. html). Published 26.01.2017, retrieved 17.04.2020.

Mbwiga, G. 2018. "Tanzania Admits BRT Project Blunder". Accessed at *The Citizen Tanzania* (http://www.thecitizen.co.tz/News/Tanzania-admits-BRT-project-blunder/1840340-4577630-6w1sbdz/index.html). Published 24.05.2018, retrieved 17.04.2020.

Meyer, M. 2012. "Placing and Tracing Absence: A Material Culture of the Immaterial". *Journal of Material Culture* 17:1, 103–110.

Michael, C. 2018. "Wataka daladala badala ya Mwendokasi". Accessed at *Mwananchi* (http://www.mwananchi.co.tz/habari/Wataka-daladala-badala-ya-mwendokasi/1597578-4822182-pxq5tcz/index.html). Published 25.10.2018, retrieved 17.04.2020.

Mngodo, E. 2016. "DART Service Off to Chaotic Start". Accessed at *The Citizen Tanzania* (http://www.thecitizen.co.tz/News/Dart-service-off-to-chaotic-start-/1840340-3206540-wj52bk/index.html). Published 17.05.2016, retrieved 17.04.2020.

Mwangonde, H. 2015. "No End in Sight for Dart Project". Accessed at *The Citizen Tanzania* (http://www.thecitizen.co.tz/News/No-end-in-sight-for-Dart-project/-/ 1840340/2701858/-/oeb1fv/-/index.html). Published 30.04.2015, retrieved 17.04.2020.

Mwisaleo 2015. "Waziri Hawa Ghasia azindua mafunzo ya madereva wa mabasi ya Mwendo wa Haraka Dar es Salaam leo". Accessed at *Mwisaleo* (http://mwisaleo. blogspot.com/2015/08/waziri-hawa-ghasia-azindua-mafunzo–ya.html). Published 17.08.2015, retrieved 11.11.2018.

Nachilongo H. 2017. "BRT Facility to Feed Off Part of Sh660bn Metro Funding". Accessed at *The Citizen Tanzania* (http://www.thecitizen.co.tz/News/BRT-facility-to-feed-off-part-of-Sh660bn-metro-funding/1840340-4200362-12otnns/index. html). Published 24.11.2017, retrieved 17.04.2020.

Namkwahe, J. 2019. "Public Safety Concern as Dart Buses Lack Emergency Kits". Accessed at *The Citizen Tanzania* (https://www.thecitizen.co.tz/news/Public-safety-concern-as-Dart-buses-lack-emergency-kits/1840340-5359342-10f2rdgz/index.html). Published 23.11.2019, retrieved 17.04.2020.

Peck, J. and N. Theodore 2010. "Mobilizing Policy. Models, Methods, and Mutations". *Geoforum* 41, 169–174.

Pineda, A. 2011. "The Map of Transmilenio. Representation, System, and City". *STS Encounters* 4:2, 79–110.

Pineda, A. 2010. "How Do We Co-Produce Urban Transport Systems and the City? The Case of Transmilenio and Bogotá". In: I. Farías and T. Bender (eds.), *Urban Assemblages. How Actor-Network Theory Changes Urban Studies*. Oxon, New York: Routledge, 123–138.

Star, S. 1999. "The Ethnography of Infrastructure". *American Behavioral Scientist* 43:3, 377–391.

The Citizen Reporter 2020. "The Worst Yet to Come, Says TMA". Accessed at *The Citizen Tanzania* (https://www.thecitizen.co.tz/news/-Rains--The-worst-yet-to-come--says-TMA/1840340-5522786-12snmly/index.html). Published 13.04.2020, retrieved 17.04.2020.

The Citizen Reporter 2018a. "Dar's Rapid Buses Transport Operator Suspends Services to City Centre". Accessed at *The Citizen Tanzania* (http://www.thecitizen.co.tz/News/Udart-suspends-services-to-City-Centre/1840340-4394148-x2ehpv/index.html). Published 15.04.2018, retrieved 17.04.2020.

The Citizen Reporter 2018b. "If You're in Dar es Salaam, Don't Try to Travel to these Places Today!" Accessed at *The Citizen Tanzania* (http://www.thecitizen.co.tz/News/1840340-4395672-36xchsz/index.html). Published 16.04.2018, retrieved 17.04.2020.

The Citizen Tanzania 2016. "Gleaming New Buses Challenge Chaotic Old Ways in Tanzania". Accessed at *The Citizen Tanzania* (http://www.thecitizen.co.tz/News/Business/Gleaming-new-buses-challenge-chaotic-old-ways-in-Tanzania/1840414-3355910-cpkcr7/index.html). Published 24.08.2016, retrieved 17.04.2020.

Thrift, N. and J. May 2001. "Introduction". In: N. Thrift and J. May (eds.), *Timespace. Geographies of Temporality*. London, New York: Routledge, 1–46.

UDA 2017. *UDA Leo. Zama Mpya*. Toleo na. 05, Mei 2017.

Ureta, S. 2015. *Assembling Policy. Transantiago, Human Devices, and the Dream of a World-Class Society*. Cambridge, MA: MIT Press.

Ureta, S. 2014. "Normalizing Transantiago. On the Challenges (and Limits) of Repairing Infrastructures". *Social Studies of Science* 44:3, 368–392.

Ureta, S. 2012. "Waiting for the Barbarians. Disciplinary Devices on Metro de Santiago". *Organisations* 20:4, 596–614.

Von Schnitzler, A. 2008. "Citizenship Prepaid: Water, Calculability, and Techno-Politics in South Africa". *Journal of Southern African Studies* 34:4, 899–917.

Winner, L. 1980. "Do Artifacts Have Politics?" *Daedalus* 109:1, 121–136.

Wood, A. 2015. "Multiple Temporalities of Policy Circulation. Gradual, Repetitive and Delayed Processes of BRT Adoption in South African Cities". *International Journal of Urban and Regional Research* 39:3, 1–13.

World Bank 2017. "Projects and Operations. Project. Dar es Salaam Urban Transport Improvement Project". Accessed at *The World Bank* (http://projects.worldbank.org/P150937?lang=en). Published 08.03.2017, retrieved 17.04.2020.

Wright, L. and W. Hook 2007. *Bus Rapid Transit Planning Guide*. New York: ITDP.

7 To conclude

Assembling BRT in the Global South

DART (Dar es Salaam Rapid Transit) is special in many ways: it is the first high-quality bus rapid transit (BRT) system in a low-income country, the first BRT system in East Africa, the first large-scale public-private partnership (PPP) in Tanzania, and one of the biggest infrastructure projects in Dar es Salaam. The processes of planning, implementing and operating DART have reshaped not only the public transport sector and everyday practices in Dar es Salaam, but also the country's political landscape and its role in the international development sector. Diverse actors have adapted to the new setting by either following or rejecting the new logics, patterns and rules that DART has brought to the Tanzanian metropolis. At the final stage of constructing the BRT infrastructure, it was hard for me to imagine how DART buses would run through the city centre, and how daladala and other vehicles, street vendors and pedestrians would share the space with the big sky-blue buses. Only a little over a year later, in May 2016, I saw them flowing along Morogoro and Kawawa Road, leaving behind daladala and other vehicles stuck in traffic, and hooting to clear their way on the busy Msimbazi Street (see Figure 7.1).

Whether the project can be called a success is not clear – and depends on one's perspective. During rush hour, DART is many times faster than daladala, at least in theory. However, since the transport system's inauguration in May 2016, several essential conditions for offering fast, reliable and comfortable bus services have been unstable or intermittent. The corridor needs to be accessible during periods of heavy rain, buses need to be maintained and repaired, drivers' salaries need to be regularly paid, passengers need to be able to buy tickets even if they do not have the exact fare. Several incidents have led to such serious operational challenges that the Tanzanian President himself issued a stern warning to the DART Agency's CEO in 2018 (Sauwa and Elias 2018). Most controversial for DART's assembling has been the biannual inundation of the BRT corridor and the bus depot at Jangwani, which can interrupt DART services for several days. In addition, the shortage of buses is a perennial topic, which was illustrated in a photo showing passengers being pushed through bus windows so that they could secure seats. DART's stakeholders were worried that the image might go

Figure 7.1 Street vendors and pedestrians along the DART corridor (Ntevi 09/2016).

global. The service delivery manager of UDA-RT wrote in the stakeholders' WhatsApp group in October 2018:

> Let us not lose the image of BRT … Such a picture can be used inappropriately and distributed WORLDWIDE and will harm the GOOD REPUTATION that has been constructed. LET US OBSERVE THIS WITH EYES WIDE OPEN so that it [the image] does not become PUBLIC…. […] it has become like a MODERN DALADALA instead of BRT.. this must be stopped VERY FAST … IT SPOILS the reputation and ITS INTENDED meaning …. At the end of the day WAZUNGU [the Westerners/Europeans/white people] WILL SAY THE BLACKS We are incapable SOMETHING THAT IS NOT TRUE …NOT EVEN A BIT.[1]

The picture went viral anyway, appearing on the front pages of most Tanzanian newspapers the next day. Two weeks later, the DART Agency published a statement in which it declared:

> The DART Agency will guarantee to continually improve the bus services by having a schedule that corresponds to the needs of the customers, and by increasing the number of buses in order to hasten the advent of a second service provider […] and ensuring that services are provided as originally planned.[2]

(DART Agency 2018b)

Nevertheless, DART has remained operational under these challenging conditions (Malanga 2020). For the time being, people have found ways to deal with the delayed and gradual implementation, accompanied by political tension and economic insecurity. In addition, BRT proponents spare no effort to project an image of a successful DART system to other African cities and to the global BRT community.

This book has sought to contribute to discussions on the growing phenomenon of mobile policies and global planning. It explored in detail how a travelling model is translated to a specific context, and how infrastructure politics materialise in a transport project. I focused on often unconsidered moments occurring between, during and after a model's mobilisation and implementation, and asked: How do global policies mutate on the move, and how are they assembled in a specific locale? DART's assembling enabled me to trace typical translational processes of travelling models. A model becomes political through its narratives, stories and discourses on the one hand, and its materialisation and adaptation on the other. DART's inscribed need to succeed, and to become the new best practice model of BRT for African cities, has become a self-fulfilling prophecy. The project has received a great deal of attention from transport planners and politicians worldwide, since 'Africa is the new Mecca of urban transport', as DART's chief technical advisor adequately described (HM 03/2015). The global BRT community is waiting and watching, curious to see if promises of BRT as 'democracy in action' (Peñalosa 2013) and 'more than bus lanes' (ITDP 2018b) are fulfilled in Dar es Salaam.

Structured according to the five main chapters of the book, this chapter reflects on the major findings of translating global models and materialising infrastructure politics. Contributing to literatures of STS and policy mobilities, it shows the advantages of a gathering approach for investigating infrastructural processes and a double mobility approach for researching transport infrastructures. The chapter summarises the role of global consultants, the fluidity of travelling models and the notion of an 'African BRT', and it elucidates the permanent restructuring of DART through both the non-compliance of in-between actors as well as through everyday practices.

Gathering in space and time

Inspired by Latour (2004) and Law (2004), my approach has been a form of gathering. It considers the temporalities and spatialities of global BRT and DART on the one hand, and the heterogeneity of assemblages on the other hand. Focusing on the processuality of the transportation project, I closely followed the implementation process of DART in Dar es Salaam in 2015 and 2016, beginning field research before it emerged that the project had indeed become the new best practice of global BRT. I have also gathered DART from documents, stories and conversations, both before and after fieldwork, and followed global policy makers in their attempts to disseminate the story of DART to Africa and the world. I closely worked with a consultancy for several months, another novelty in policy mobilities studies.

Following Law's thoughts on method assemblage (2004), I adapted myself to the topic of research by remaining mobile and flexible, which helped me to notice unexpected developments and details. Using a double mobility approach, I explored DART, the transport system, by moving with it along its city corridor; and I explored DART, the globally travelling model, by following its connections to an office in Nairobi, to narratives originating in Bogotá and elsewhere. Considering both the technical and political aspects of planning and how they are interlinked, I interacted with planners, politicians, consultants and transport operators. I also investigated materials and things; in Dar es Salaam, buses and the automated fare-collection system (AFCS) as well as speed bumps and fences contributed considerably to the system's becoming. Their material presence and their materialities themselves enabled the disciplining of road users and the strategy of non-compliance (see below). Hence, the non-negotiable character of materials and things had a strong impact on DART that challenged written words of contracts and signs (c.f. Latour 1994, 1999). Building upon these empirical approaches and findings, I demonstrated how flexible and mutable a global model must be to materialise in the local, and how standardised and immutable it must be to remain the same across contexts.

Fluid formations of global transport

Disseminating and implementing BRT involves two intertwined processes that might seem contradictory at a first glance: the model needs to be stabilised through standardisation, but also by keeping a certain degree of flexibility, i.e. adaptability. Hence, global consultants have been striving to standardise the global transport model, their efforts materialised in guides, webinars and awards, among other things. They also aim to increase the unifying features of BRT systems worldwide by decreasing the model's mutability. Notwithstanding recent standardisation attempts, the translation of the BRT model to different contexts not only involves displacement and adaptation but, above all, implies mutation. For instance, ITDP's *BRT Standard* – a self-referential tool that defines BRT's characteristics – is itself constantly changing through periodic updates (see ITDP 2016). According to Callon's (1986) conceptualisation of translation, mobilisation is only possible through the model's inherent flexibility – i.e. its fluidity and mutability. This complex translation work requires a wide variety of actors. In addition to global consultants from ITDP and the World Bank, Tanzanian politicians, Dar es Salaam's passengers, the daladala sector and the network of UDA-RT, nonhumans such as buses and concrete, rain and heat, guides and standards, regulations and laws, paper tickets and smart cards, have been taking part in DART's assembling.

Globalising a model requires not only mobilisation but also localisation. The existence of the global model and its materialisations in different locales are interdependent; they mutually influence and shape one another. Following Behrends et al.'s (2014) argument that travelling models are an outcome of global assemblages (Ong and Collier 2005), I argue that each assembling of a BRT system is influenced by the travelling model and, because the model itself is not rigid, that each individual system may also feed back into the travelling model. To what extent DART shapes the global BRT model has yet to be determined – but it is clear that DART impacts the global model through its presence in ITDP's recent publications and other activities, which have created the 'African version' of best-practice BRT. Actors in this field have experienced both adaptation and mutation, and hence they try to locate the travelling BRT model between standardisation and flexibility. This double endeavour is partially reflected in ITDP's *BRT Standard*, with its accreditation system ranging from *BRT Basic* to *BRT Gold*. However, ITDP works hard to convey the impression that mutation does not happen during the process of translation – which leads to a contradiction between the image of BRT as a coherent, high-quality public transport model on the one hand, and the diverse, sometimes controversial ways that BRT is actually implemented on the other. Consultants downplay deviation and mutation of plans and models because these unexpected and only partially controllable processes could indicate controversies, political conflicts and public disagreement. Consultants fear that mutation might destabilise not only the specific model itself but also the general idea of models themselves. However, the reverse is the case: if BRT consultants acknowledge that mutation is inherent in travelling models and not necessarily a risk, they can prevent the friction and uncertainty that emerges when a BRT system is not exactly implemented as it was planned and advertised. Furthermore, planners and politicians could be prepared to react appropriately to the fluidity of a project because flexible plans facilitate a system's adaptability to a specific context. This implies taking (in)coherence, ambivalence and multiplicity into account (Hine 2007). If planners and consultants let practice co-determine the plans, the project could remain accessible to contributions by new actors. However, neither strictness nor openness guarantee smooth implementation. Complications and delays can and do occur in any planning process, as when UDA-RT ordered more buses than officially agreed.

During its implementation, DART's plans have been changed and renegotiated several times; the planning process has been able to respond to practice because it has remained flexible. Hence, assembling DART has meant deviating from the global BRT model and the initial operational designs of DART. I thus refer to DART's assembling as a fluid formation. Drawing on debates on the fluidity of technologies (De Laet and Mol 2000), the mutability of mobilised things and policies (Latour 1986; McCann and Ward 2013), this research has shown that mutation of travelling models is more the rule than the exception, and indeed that it is indispensable in the localisation of the model. However, all travelling models need a strong centre to hold together their essential characteristics. For instance, dedicated lanes and off-board fare collection systems are indispensable for a materialised BRT system (but the model offers variations of those characteristics in terms of size, materials and technologies used, etc.). BRT is more mutable than Latour's (1986, 1987) printing press; its fluidity is similar to De Laet and Mol's (2000) bush pump. Therefore, understanding DART as a fluid formation means recognising that the transport system consists of heterogeneous actors, and that its process of becoming is characterised by mutation.

The model of the Global South: frictions and success

In contrast to other infrastructure systems, BRT has heroic figures and a clear plotline (see Bowker 2015) – two ingredients that facilitate the global circulation of both the BRT model itself as well as the stories of individual systems like DART. The global BRT community is united in trying to create a successful image of BRT, as this success reflects the work consultants and technocrats have invested. This, however, makes the individual actors mutually dependent: ITDP needs funding from (development) banks, foundations and development cooperation agencies, which in turn need the expertise of ITDP to realise the BRT projects. Meanwhile, local and national governments are keen to receive funding for future phases of DART or other infrastructure projects. Thus, everyone uses DART to 'validate' their own work and to demonstrate that BRT is indeed the best and only option for future urban public transport in the Global South. Technocrats' strategic practices of dissemination and politics of persuasion – like storytelling and sugar-coating – influence the creation and mobilisation of the BRT model. Politicians, consultants and scholars draw on religious vocabulary to describe the BRT universe, and they put different emphases on either social and ethical (ITDP and Peñalosa) or economic (World Bank) dimensions. The notion of success is a key ingredient for the model's rapid dissemination in the Global South, which politicians like Magufuli might productively use for their political endeavours. Moreover, global policy models are often associated with specific sites, and they carry labels, which enable them to become mobile, famous and thus successful. BRT stands for Bogotá and carries the label of being the only affordable and sustainable public transport solution for cities of the Global South. In global representations of BRT, Bogotá's 'success' is omnipresent, and Delhi's 'failure' is absent. Also, the Global North is absent in BRT narratives: just as *Transmilenio* served as best practice for Dar es Salaam and Jakarta, DART is now serving as best practice for Kampala and Addis Ababa. On the assumption that Southern metropolises have comparable transport needs and similar spatial and institutional conditions, BRT facilitates new connections across these cities.

However, the label 'model of the Global South' evokes three intertwined frictions that indicate the need for a critical perspective on policy mobilities – one that considers historical difference and global hegemonies, and that is open to relational thinking and Southern theorisations (Roy 2015). The first friction reflects a general critique on the adequacy of categories like 'South' and 'Africa'. Particularly from a white, privileged standpoint like my own, these categories tend to divide the world into territorial entities and reduce the role of relational thinking. Furthermore, these categorisations homogenise the Global South, which is characterised by heterogeneity, asymmetries and hegemonies. While there are some similarities between cities like Bogotá, Jakarta, Dar es Salaam and Kampala – the common initial situation (rapid urban growth) with the same mode of public transport (minibus systems) and a similar financial background (dependent on foreign investment and very limited governmental subsidy for public transport) – numerous aspects are significantly different, especially institutional settings, the internal structure of the minibus sectors and local protest against BRT. Therefore, how can DART exemplify the future of African transport?

The second friction relates to the presumed smooth policy transfer from site to site, in this case from South to South, by replicating models and experiences, which does not conform to actual experiences of translating policy models like BRT. In Dar es Salaam, the physical as well as the operational design – both heavily inspired by *Transmilenio* – led to tremendous delays arising from disagreement and protest. While translating physical designs went relatively smoothly, the operational design has mutated a great deal – a fact that BRT consultants downplay when they present DART to the global BRT community. Translation is much more complex, uncertain and contingent than global consultants make it out to be.

The third friction concerns the global hegemonies and inherent interdependencies of BRT circulation. This tight network of BRT consultants that Rizzo (2017) aptly named 'the Evangelical Society' brings the same few protagonists to every Southern BRT context. This network consists of individuals and institutions that are mainly based in and connected to North America and Europe. ITDP and the World Bank are at the forefront, often accompanied by well-established development cooperation agencies and bus manufacturers. For instance, the work of ITDP – which presents itself as an independent consultancy – is highly strategic and normative insofar as the NGO also acts as an (unofficial) BRT proponent. Because they create knowledge and define expertise, these 'Northern entities with a Southern eye' have extraterritorial authority (Wood 2014) and 'circulatory capacity' (Roy 2012). The protagonists of the BRT community influence the futures of cities of the Global South by forming opinions on preferable means of transportation, financial models and the use of land. They also decide to what extent local politicians, planners and transport operators become involved in the transition of urban public transport, although they are not always assertive, as the actions of UDA-RT showed.

Effective operational controversies: non-compliance as resistance

The most controversial aspect of implementing BRT in Dar es Salaam has been the operational model. Within this controversy, not all transport actors have played their allocated roles and met international conditions of the project. Rather, certain local operators have taken advantage of the common interest of the World Bank, ITDP and the Tanzanian government in the project succeeding at all costs to impose their own idea of DART's operational structure. The existence of DART has never been put into question as such, so it has never risked being abandoned like BRT in Delhi or the private rapid transit project *Aramis* in Paris (Latour 1996). Support in Dar es Salaam has, however, not been unanimous from the Tanzanian government or the transport business. Connections between government officials and UDA, the largest public transport company, raised questions regarding the participatory structure of the policy model. UDA-RT's claims to participate in and benefit from the new infrastructure made DART's assembling highly controversial. The various changes, massive delays and general difficulties regarding the physical infrastructure and operational design have been based on controversies that involved negotiations of power and participation, expressed most strongly in relationship to DART's operational design.

Contributing to the policy mobilities literature, this research has shown that a closer examination of the assembling at a specific site sheds light on the global circulation of policies and power, models and expertise, and the inherent tensions and frictions involved. Even though mutation is an essential part of translating policy models and not necessarily problematic for the actors of a project, DART's mutations have fundamentally influenced the system's assembling. Adaptation is not as smooth as global consultants proclaim, and hence it is only made visible by looking closer at the formation and mutation of travelling models. Often, research on travelling policies attaches little importance to processes of adaptation and translation, instead focusing on global relationships and movement. In this research, I followed policies and models while they were travelling and during their assembling. Present absences of material components like number plates, logos and smart cards, as well as the legal structure of the bus operator, changed repeatedly. Controversies are present as they occur and made absent quickly in the aftermath so that a research from a retrospective angle would have difficulties in identifying them. This approach brought meaningful insights into heated debates and negotiations about who participates in the new transport system.

Inspired by Callon et al.'s understanding of a controversy as a 'highly effective apparatus' that 'enrich[es] the meaning of a situation' (2009: 30), I dubbed the strategies and power games of DART's operational design 'effective operational controversies'. I thereby focused on the controversies' potential to explore alternatives, (re)open space for negotiation and provide opportunity for a transformation of the project. This perspective goes beyond seeing mobile policies as either a success or a failure – a viewpoint that persists in research and practice (Lovell 2017; McCann and Ward 2015). Instead, flexible planning and an openness to learning from failures can facilitate a smooth adaptation of global models, whereby one-sided success stories might create a misleading image. The effects of these powerful controversies are unpredictable; they can be productive or destructive. Even if operational controversies enabled transport operators in Dar es Salaam to secure roles in the project, the effects of these controversies harmed almost everyone in financial or reputational terms: the World Bank questioned its future investment in DART, ITDP and the Tanzanian government feared a loss of DART's image, UDA-RT's workers did not receive work contracts, and passengers had to deal with unreliable services.

The controversies over DART revealed two main features that differed from the common picture presented by research on public, scientific or knowledge controversies (Barry 2012; Jasanoff 2012; Nelkin 1979). First, the actors who evoked the controversy were neither laypersons like citizens and transport workers, nor experts like high-ranking politicians and international NGOs. In-between actors were the main opponents: UDA and UDA-RT, indirectly supported by government elites. Investigating intermingled power structures and complex networks of Tanzania's political economy explained how these actors in between have succeeded in enforcing their operational participation and profit-sharing, thereby contesting global best practice. Second, the main strategies of local transport operators – merging the two leading transport actors in order to ruin the idea of competition and investing a great deal of money in technology to make themselves irreplaceable – do not count as typical or usual forms of resistance; neither against a new technology, nor against a BRT project.

I dubbed UDA-RT's strategy to become and remain a permanent powerful actor of DART 'non-compliance as resistance'. The company deployed a proven strategy that had been successful in the past: disregarding contracts with the government and making massive investments without official authorisation. Moreover, UDA-RT applied subtle strategies to oppose and outwit international agreements, e.g. by excluding non-Swahili speaking consultants, undertaking last minute changes to service provider contracts, and taking over the headquarters of the DART system. Considering non-compliance as a possible mode of mutation and – thus as an inherent part of translation (Callon 1986) – UDA-RT's strategies can be regarded as a step that was to be expected. Because the company has succeeded in securing and maintaining its participation in DART, its forms of non-compliance constitute a successful translation of global models into context-specific circumstances that fulfil local expectations. The initial operational model for DART included two competing bus operators, each of which was a coalition of a national and an international company. When local transport elites contravened this international opening of the sector, the DART model transformed into the UDA-RT model, which gives a minimal role to the state and shows significant similarities to the operational model of daladala. Crucial for discouraging international companies from becoming part of DART is non-transparent communication, strong relationships with governmental elites for indirect support, the preparedness to make extensive financial investments, and a general unwillingness to comply with official agreements.

Permanent interim assemblages: (not) a modern daladala

To emphasise the various temporalities, heterogeneities and contingencies of a travelling model's assembling, this work described DART as a permanent interim assemblage. DART's continuous and extended implementation process has made it permanently interim. The transport system was inaugurated in 2016, but the project is about the same age as the global success story of BRT; both began development in the early 2000s. This ethnography has contributed to discussions of fast global policies and their slow materialisations (see Peck and Theodore 2015; Wood 2015) by explaining the prolonged and interim character of planning and implementing DART. When researching processes of policy circulation, the focus on speed has only limited utility. Instead, examining permanently occurring mutations reveals controversies and gives socio-political meaning to research endeavours. As DART's non-linear and controversial assembling has shown, the system's continued existence between global ideals, initial plans and current practices demonstrates the tensions of global transport planning and the contradictions between global narratives and local practices. In terms of the intermittent assembling of DART, my empirical approaches 'gathering' and 'double mobility' have revealed global relations and interruptions, as well as loops, standstills and accelerations of the process. Just as with controversies, these temporalities would have been impossible to discover retrospectively.

The assembling of DART has been characterised by a combination of premature aging and deferred repair and maintenance, which has led to the rapid deterioration of the physical infrastructure. On the one hand, provisional material assemblings have stabilised so that half-built fences and loose ropes have become permanent within terminals, and passengers stopped hoping for additional smart cards. On the other hand, the system is gradually improving and continuously being set up. The control centre is becoming more professional, real estate development is increasing along the corridor, one additional feeder route has been opened, and DART frequently hosts international delegations. DART's provisional operations, lacking essential BRT components, remain in a permanently interim status. This is most clearly expressed in the interim service provider (ISP) quietly dropping the 'I' to become the (permanent) service provider (SP). However, this modification has not been reflected in DART's structures, and operations have not changed accordingly, so that it remains uncertain whether a competing bus operator will be added.

On the ground, DART blends daladala practices with global BRT ideals. Its assembling thus wavers between the labels 'modern daladala' and 'world-class BRT'. Routine overcrowding and incidents like inundation have led to serious operational challenges and revealed that DART resembles daladala both in terms of its route network and its operational aspects, such as the absence of schedules. Most of DART's passengers and operators do not problematise the limited use or the deferred implementation of the intelligent transport system (ITS) and the AFCS. For instance, passengers do not expect ITS components like signs and displays at stations or in buses but instead translate daladala practices into DART's operations, waiting for announcements from the driver via speakers and asking fellow passengers about the route or the final destination of a bus. Intermittent assembling thus means being able to put plans into practice by modifying them according to local conditions and needs.

Even if the new transport system in Dar es Salaam does not really reflect the BRT ideals of high quality, comfort and safety, the global BRT community has managed through hard, persuasive work that DART is being recognised globally as high-quality BRT system. This permanent interim assemblage may have clear advantages because its inherent heterogeneity, temporalities and contingency also include the opportunity to flexibly react to how the model is locally adapted.

DART as the African model?

Sometimes, I was afraid that DART would not eventuate. DART, like *Aramis*, has been afflicted with the lack of a strong 'local engine' (Latour 1996: 85), but it differs from the French PRT (personal rapid transit) on at least three key points. First, DART has always had actors on different scales that showed clear enthusiasm. Second, assembling DART goes back to a relatively stabilised model that refers to experiences from elsewhere. Third, the global BRT model is defined by global consultants who generally acknowledge that circulating transport models involves mutation. Planners and politicians did not separate the technical from the social of DART, but instead responded to DART's need to be flexible. The option to negotiate and to adapt to the context of Dar es Salaam allowed DART to begin operations, even while controversies continued to rage. However, DART's operations created a discrepancy between global narratives of success and local voices of critique. Despite DART's similarities with daladala and its fundamental deviations from the global BRT model, interdependent actors emphasised that DART is 'a no-regret investment' (BN 03/2015), and that BRT was the only solution for the city to remain functional and to meet everyone's needs (HM 03/2015; ITDP 2018a). DART's extended implementation of Phase 1 exceeded the maximum time planned for stabilisation, particularly due to several unsuccessful attempts to find a second bus operator. The blame game and power struggles between the Tanzanian government and the private transport operator are ongoing, and ITDP is still hesitating to officially score this supposed new best practice for the *BRT Standard*. Maybe the NGO is waiting for fundamental improvements in Dar es Salaam so that its 'Golden forecast' will still come true?

That DART has not been cancelled along the way is predominantly due to the global interdependencies that unify competing actors because they pursue the same overall objective: for these actors, DART must succeed and become the African BRT model. These interdependencies reflect two intermingled political moments of pressure and opportunity. First, the global BRT community urged everyone involved to strive so that the 'guinea pig' (HM 03/2015) indeed becomes the 'watershed moment' for African cities (ITDP 2018c). Second, ITDP's and the World Bank's early initiative, permanently active role and clear aims have opened broad possibilities to Dar es Salaam and Tanzania's global stance in terms of future investments. In addition to the obvious and primary changes in Dar es Salaam's transport system, infrastructure projects like DART, flyovers and bridges have had a spillover that materialises in new economic investments. Despite achieving high ridership, shortening average travel times and reducing carbon emissions per passenger, DART's mission to be an accessible public transport system that enables poverty reduction, improved standards of living and sustainable economic growth has yet to be fulfilled (see DART Agency 2018a). Everything depends on future operations: Will UDA-RT ever get a competitor in Phase 1? When will Phase 2 be operational, and by whom? How will the World Bank reshape its conditions to guarantee compliance in Phases 3 and 4? Will Phases 5 and 6 be constructed as initially planned, or will Japan International Cooperation Agency's (JICA) plan to construct rail finally win out?

Despite DART's uncertain future, it is clear that this fluid transport technology will keep changing during further construction and operation. DART's combined correspondence to and deviation from the global BRT model raise three unanswered questions: To what extent will the Tanzanian version of BRT influence the global model? Has DART deviated more from the global model and initial plans than other BRT projects? And can DART cope with its role as the African BRT model, and establish itself as the new best practice in the long run? For global BRT consultants like ITDP, DART has proved to be an effective way to gain a foothold on the African continent, and to disseminate the BRT model – which stands for high-quality operations in low-income countries – to various African cities. Alternatively, for local transport operators in cities of the Global South, the UDA-RT model might better serve them as best practice because it enables them to take over the new transport infrastructure, despite global financiers' requirement for an international public-private partnership (PPP) model: non-compliance as a form of resistance could become an expression of self-empowerment and emancipation. But how to disseminate the UDA-RT model and make it a global best practice of its own is not (yet) clear. Not only is the political context of UDA-RT very specific, which would require the model to be highly mutable, local transport actors are also not well connected globally and would find it difficult to do persuasive work elsewhere. Since the technopolitical dimensions of infrastructure planning and the translation of models – BRT, DART and UDA-RT – are mutually dependent, the (further) mobilisation of these models in the Global South depends on the interests of BRT proponents and local transport actors. The globally circulating transport model and the systems under implementation and operation always include technologies, politics and societies. DART, situated between evangelical narratives of transformation and success and the ambivalences of enthusiasm and controversy in Dar es Salaam itself, has shown that BRT is indeed far more than just a transport system.

Notes

1 *Tusipoteze Image ya BRT ... Picha kama hii inaweza kutumika vibaya na kusambaa DUNIANI na kuondoa SIFA KUBWA liyojengeka. TULITAZAME KWA JICHO PANA hili ili lisiweke DOA.... [...] inakuwa kama MODERN DALADALA badala ya BRT .. hii lkemewe kwa HARAKA SANA .... INAPOTEZ sifa na Maana ILIYOKUSUDIWA .... Mwisho wa Siku WAZUNGU WATASEMA WEUSI Hatuwezi JAMBO AMBALO SI KWELI ...HATA KIDOGO.*
2 *Wakala ya Mabasi Yaendayo Haraka itahakikisha siku zote kuwa huduma za mabasi zinaboreshwa kwa kuwa na ratiba inayozingatia mahitaji ya wateja na kuongeza idadi ya mabasi kwa kuharakisha hatua za kumpata mtoa huduma wa pili [...] na kuhakikisha kwamba huduma zinatolewa kulingana na ratiba iliyopangwa.*

References

Barry, A. 2012. "Political Situations. Knowledge Controversies in Transnational Governance". *Critical Policy Studies* 6:3, 324–336.

Behrends, A., S. J. Park and R. Rottenburg 2014. "Travelling Models. Introducing an Analytical Concept to Globalisation Studies". In: ibid. (eds.), *Travelling Models in African Conflict Management. Translating Technologies of Social Ordering.* Leiden: Brill, 1–40.

Bowker, G. 2015. "Temporality". In: H. Appel, N. Anand and A. Gupta (eds.), *The Infrastructure Toolbox.* Fieldsights, September 24. Accessed at *Cultural Anthropology* (https://culanth.org/fieldsights/series/the-infrastructure-toolbox). Published 24.09.2015, retrieved 17.04.2020.

Callon, M., P. Lascoumes and Y. Barthe 2009 (2001). *Acting in an Uncertain World. An Essay on Technical Democracy.* Cambridge, MA: MIT Press.

Callon, M. 1986. "Some Elements of a Sociology of Translation. Domestication of the Scallops and the Fishermen of St Brieuc Bay". In: Law, J. (ed.). *Power, Action and Belief. A New Sociology of Knowledge?* London: Routledge, 196–223.

DART Agency 2018a. "DART. Dar Rapid Transit". Accessed at *DART Agency* (http://www.dart.go.tz/publications/2). Published October 2018, retrieved 11.11.2018.

DART Agency 2018b. "Taarifa kwa Umma". Accessed at *DART Agency* (http://www.dart.go.tz/). Published 22.10.2018, retrieved 11.11.2018.

De Laet, M. and A. Mol 2000. "The Zimbabwe Bush Pump. Mechanics of a Fluid Technology". *Social Studies of Science* 30:2, 225–263.

Hine, C. 2007. "Multi-Sited Ethnography as a Middle Range Methodology for Contemporary STS". *Science, Technology & Human Values* 32:6, 652–671.

ITDP 2018a. "BRT Planning 101". Accessed at *ITDP* (https://www.itdp.org/publication/webinar-brt-planning-101/). Published 02.02.2018, retrieved 17.04.2020.

ITDP 2018b. "BRT Planning 501: Integration". Accessed at *ITDP* https://www.itdp.org/2018/10/09/webinar-brt-501-integration/). Published 09.10.2018, retrieved 17.04.2020.

ITDP 2018c. "Dar es Salaam, Tanzania Presented with 2018 Sustainable Transport Award". Accessed at *ITDP* (https://www.itdp.org/2018-sta-ceremony/). Published 30.01.2018, retrieved 17.04.2020.

ITDP 2016. "BRT Standard. 2016 Edition". Accessed at *ITDP* (https://www.itdp.org/2016/06/21/the-brt-standard/). Published 21.06.2016, retrieved 17.04.2020.

Jasanoff, S. 2012. "Genealogies of STS". *Social Studies of Science* 42:3, 1–7.

Latour, B. 2004. "Why Has the Critique Run Out of Steam? From Matters of Fact to Matters of Concern". *Critical Inquiry* 30, 225–248.

Latour, B. 1999. *Pandora's Hope. Essays on the Reality of Science Studies.* Cambridge, MA: Harvard University Press.

Latour, B. 1996. *Aramis or the Love of Technology.* Cambridge, MA: Harvard University Press.

Latour, B. 1994: "On Technical Mediation". *Common Knowledge* 3:2, 29–64.

Latour, B. 1987. *Science in Action. How to Follow Scientists and Engineers through Society.* Cambridge, MA: Harvard University Press.

Latour, B. 1986. "Visualisation and Cognition. Drawing Things Together". *Knowledge and Society – Studies in the Sociology of Culture Past and Present* 6, 1–40.

Law, J. 2004. *After Method. Mess in Social Science Research.* London, New York: Routledge.

Lovell, H. 2017. "Policy Failure Mobilities". *Progress in Human Geography* 43:1, 1–18.

Malanga, A. 2020. "Fresh Move to Sort Out Dart Chaos". Accessed at *The Citizen Tanzania* (https://www.thecitizen.co.tz/news/1840340-5462526-af1mr4/index.html). Published 20.02.2020, retrieved 17.04.2020.

McCann, E. and K. Ward 2015. "Thinking Through Dualisms in Urban Policy Mobilities". *International Journal of Urban and Regional Research* 39:4, 828–830.

McCann, E. and K. Ward 2013. "A Multi-Disciplinary Approach to Policy Transfer Research. Geographies, Assemblages, Mobilities and Mutations". *Policy Studies* 34:1, 2–18.

Nelkin, D. 1979. *Controversy: Politics of Technical Decisions.* Beverly Hills: Sage.

Ong, A. and S. Collier (eds.) 2005. *Global Assemblages. Technology, Politics, and Ethics as Anthropological Problems.* Malden, MA: Blackwell.

Peck, J. and N. Theodore 2015. *Fast Policy. Experimental Statecraft at the Thresholds of Neoliberalism.* Minneapolis, London: University of Minnesota Press.

Peñalosa, E. 2013. "Why Buses Represent Democracy in Action". Accessed at *TED* (https://www.ted.com/talks/enrique_penalosa_why_buses_represent_democracy_in_action#t-221029). Published September 2013, retrieved 17.04.2020.

Rizzo, M. 2017. *Taken for a Ride. Grounding Neoliberalism, Precarious Labour and Public Transport in an African Metropolis.* Oxford: Oxford University Press.

Roy, A. 2015. "Who's Afraid of Postcolonial Theory?" *International Journal of Urban and Regional Research* 40:1, 200–209.

Roy, A. 2012. "Ethnographic Circulations. Space – Time Relations in the Worlds of Poverty Management". *Environment and Planning A* 44:1, 31–41.

Sauwa, S. and P. Elias 2018. "Dar Transport in Chaos over Udart Troubles". Accessed at *The Citizen Tanzania* (http://www.thecitizen.co.tz/News/Dar-transport-in-chaos-over-Udart-troubles/1840340-4800784-59ig8sz/index.html). Published 11.10.2018, retrieved 17.04.2020.

Wood, A. 2015. "Multiple Temporalities of Policy Circulation. Gradual, Repetitive and Delayed Processes of BRT Adoption in South African Cities". *International Journal of Urban and Regional Research* 39:3, 1–13.

Wood, A. 2014. "Moving Policy. Global and Local Characters Circulating Bus Rapid Transit Through South African Cities". *Urban Geography* 35:8, 1238–1254.

Interviews

Additional empirical material is detailed in Chapter 2. Information on anonymisation is available in the 'Practical Notes' at the beginning of the book. The names given below are pseudonyms.

The positions indicated are those my research partners held when I interviewed them.

Key	Date	Name	Profession/Position (Institution/Company)
AE	05/2016	Astrid Engelbrecht	Customer operations manager (Austrian AFCS/ITS supplier)
AH	03/2015 10/2015 05/2016	Amir Hanif	Head (DARCOBOA), spokesperson (UDA-RT)
AK	05/2016	Alberto Kipenga	Personal secretary (Simon Group Ltd.)
AM	10/2015	Ajit Mukherjee	Bus operations specialist (DART Agency)
AP	09/2016	Ahmed Pondamali	Chief operations officer (UDA-RT)
AR	10/2015	Agata Rushaka	Regional administrative secretary (Regional Secretariat Dar es Salaam)
BN	03/2015 10/2015	Bryan Newman	Senior urban specialist (World Bank Tanzania)
CG/WU	04/2015	Chane Gumbo Willem Usisivu	Road engineer (Nairobi City County) Civil engineer (Nairobi City County)
DH/LH	10/2015	Darrell Ha Louis Hopkins	Market manager (Chinese bus manufacturer) Chief operations officer (UDA-RT)
DN	03/2015	Divit Nigam	Transport consultant (freelance, Nairobi)
DS/MH	09/2015	Detlef Schmidt Matthias Hansen	Commercial manager (Austrian road construction company) Head of technical office (Austrian road construction company)
DT	03/2015	Derek Thuku	Head (Matatu Owners Association)
EB	03/2015	Eric Bitok	Computer scientist (Digital Matatus Project) and lecturer (University of Nairobi)
EL	10/2015	Elizabeth Lugola	Head (Social Development Section, DCC)

(*Continued*)

Key	Date	Name	Profession/Position (Institution/Company)
EM/VM	09/2015	Edwin Magogo Victor Makaidi	Driving teachers (VETA, Dar es Salaam)
ER	08/2016	Ethan Robinson	Transport research and evaluation manager (ITDP, Washington DC)
FG	04/2015	Frank Gachara	Deputy chief economist (Central Planning Unit at Ministry of Transport, Nairobi)
GV	05/2016 09/2016 11/2016	Gabriel Vassanji	CEO (DART Agency)
HB/NM	09/2016	Harold Bafadhili Nelson Mtumbo	Executive chairman (Simon Group Ltd.) Managing director (Simon Group Ltd.), managing director (UDA) and board member (DART)
HG	05/2016 09/2016	Heaton Galinoma	CEO (UDA-RT)
HK	05/2016	Haruto Kobayashi	Director (Development Project Department of JICA, Dar es Salaam)
HM	03/2015 10/2015	Herbert Meyer	Chief technical advisor (DART Agency)
HM/AM	09/2015	Herbert Meyer Ajit Mukherjee	Chief technical advisor (DART Agency) Bus operations specialist (DART Agency)
JB	10/2015 09/2016	Janvier Bellamy	Financial advisor (Dutch advisory group) and leader of transaction advisory team (Dar es Salaam)
JK	03/2015 10/2015 05/2016 09/2016	James Kombe	Senior transport specialist (World Bank Tanzania)
JM	03/2015	Jason Morris	Public transport specialist (British engineering consultancy)
JO/PA	03/2015	Joseph Oboya Pamela Abira	Project manager (South African engineering consultancy) Assistant project manager (South African engineering consultancy)
JR	03/2015 10/2015 05/2016	Juma Ruete	Corporate communication manager (SUMATRA)
LH	10/2015	Louis Hopkins	Chief operations officer (UDA-RT)
LH/EI	05/2016	Louis Hopkins Edgar Ilinga	Chief operations officer (UDA-RT) Traffic officer (UDA-RT)
LL	11/2016	Luc Lavoie	Transport modeller (Brazilian engineering consultancy)
MH	09/2016	Matthias Hansen	Head of technical office (Austrian road construction company)
MN	03/2015	Mzuzi Nkutu	Project manager (TANROADS)
NM	05/2016	Nelson Mtumbo	Managing director (Simon Group Ltd.), managing director (UDA) and board member (DART)
NT	10/2015	Nicolas Thabeet	Professor (Transport Planning, University of Dar es Salaam/UDSM) and board member (DART)

Key	Date	Name	Profession/Position (Institution/Company)
OL	05/2016	Oliver Lu	Manager for Eastern Africa (Chinese bus manufacturer)
PF	10/2015	Patrick Fereji	Road engineer (TANROADS)
PH	03/2015 07/2015 11/2015 05/2016 12/2016 11/2017	Peter Hall	Africa programme director (ITDP)
RH/KD	10/2015	Rudolph Huvisa Kingston Degera	Project manager (Tanzanian engineering consultancy) Managing director (Tanzanian engineering consultancy)
RL	10/2015 05/2016	Rafael Liwanji	Director of operations and infrastructure management (DART Agency)
RM	03/2015	Rukia Malima	Project assistant (PPP Unit, Ministry of Finance, Dar es Salaam)
RO	03/2015	Ralf Oberhausen	Surveying technician (Austrian road construction company)
RP	05/2016 09/2016	Rachit Padmanabhan	Operations manager (UDA-RT)
SC	05/2016	Steven Chan	After sales manager (Chinese bus manufacturer)
SG/VM	10/2015	Salehe Gurnah Victor Makaidi	Chief instructor (VETA) Head of trucks and mechanics (VETA)
SK	10/2015	Samuel Kevela	Assistant director (Road Transport Unit, Ministry of Transport, Dar es Salaam)
TG	12/2016	Trevor Gabriel	Transport consultant and university professor (UC Berkeley)
TP/GP	05/2016	Tayla Pengo Godwin Patton	Principal legal officer (PPP Unit, Ministry of Finance, Dar es Salaam) Resident advisor (US Treasury International Affairs)
WG	10/2015	Walter Gordon	Project director (Australian engineering consultancy)
WM	03/2015 09/2016	Willfred Majaliwa	Permanent secretary (Ministry of Works, Dar es Salaam)
YK	10/2015	Yovela Kambona	CEO (DART Agency)
ZD	03/2015	Zakai Degera	Project manager (DART Agency)
ZD/GB	09/2016	Zakai Degera Garrison Buyogera	Service delivery manager (UDA-RT) Public relations officer (UDA-RT)

Index

Note: *Italic* page numbers refer to figures and page numbers followed by "n" denote endnotes.